Practical Guide to the Low Voltage Directive

Gregg Kervill

gregg@gkcl.com

Newnes

OXFORD   BOSTON   JOHANNESBURG   MELBOURNE   NEW DELHI   SINGAPORE

Newnes
An imprint of Butterworth-Heinemann
Linacre House, Jordan Hill, Oxford OX2 8DP
225 Wildwood Avenue, Woburn, MA 01801-2041
A division of Reed Educational and Professional Publishing Ltd

R A member of the Reed Elsevier plc group

First published 1998

British Library Cataloguing in Publication Data
A catalogue record for this book is available from the British Library

ISBN 07506 3745 5

Library of Congress Cataloguing in Publication Data
A catalogue record for this book is available from the Library of Congress

Typeset by ʡ Tek-Art, Croydon, Surrey
Printed in Great Britain by Biddles Ltd, Guildford and King's Lynn

Contents

About the author

Gregg Kervill delivers Product Safety lectures and consultancy services on both sides of the Atlantic and in the Pacific Rim; he has presented papers in Washington DC, Euro-EMC, London EMC and at UK EMC clubs. His book *The Practical Guide to Electrical Product Safety* and the 'Electrical Safety CD-ROM' have sold worldwide.

Gregg began life as a physicist but quickly converted to electronics, because, as he puts it, it was more fun. When pressed he will confess to twenty years of R&D experience in the high volume consumer, industrial process control, coal mining and defence industries; but will admit to no more.

He trained in Product Safety while working for Digital Equipment Company before founding GK Consultants Ltd (GKCL) in 1993.

GKCL specializes in training European companies to self-certify their products using the Manufacturer's Declaration of Conformity route of the Low Voltage Directive and to design to meet North American and Canadian safety standards. It also helps companies outside the EU to produce documentation to support CE Marking of their equipment.

Gregg Kervill	GK Consultants Ltd (GKCL)
Tel:	+44 (0) 1703 767739
Fax:	+44 (0) 1703 767789
E-mail	gregg@gkcl.com
Web	www.gkcl.com

1 Introduction

About this book

The reason for this book

This book describes the basic principles and explains, in simple terms, what the Low Voltage Directive (LVD) and harmonized standards are trying to achieve. It concentrates on how the reader can avoid the common and the more subtle non-compliances that are found in most new designs: it achieves this by taking the reader on a step-by-step tour through product design. This practical approach is what makes this book an invaluable companion for experienced engineers and a source of basic knowledge for engineering students. It answers some of the most commonly asked questions from designers and then provides detailed explanations – on a clause by clause basis – of the most widely used Harmonized Standard EN 60950.

How to use this book

How you use this book will depend primarily on your job function and experience in product safety engineering. Start by reading through and gaining a thorough understanding of the requirements described within this chapter. While reading, try to focus on the reason why rather than on the detail: only when you have a good understanding of the principles the LVD is trying to achieve will the further details become clear.

Important basics concepts

Product space versus process space – product safety versus quality

Before we can talk about or discuss any subject we must, of course, speak the same language. Within the subjects of product safety, regulatory compliance and the Low Voltage Directive we use words in special ways, and so the first task of this chapter is to explain some of the basics. Here we consider the relationship of product safety with quality.

There is a general misconception that if I buy from a quality company that I will get a fully compliant product. While this is quite a reasonable and valid assumption it is, sadly, not wholly correct. If I buy a product with the mark of a European safety agency (such as BSI, TÜV, NEMKO etc.) I can be reasonably sure that the product was tested, found to be compliant and that there will be a manufacturing audit of the product (usually) four times a year. What is more, I can have the same degree of confidence whether the manufacturer is ISO 9000 accredited or not. This immediately looks wrong because it is natural to expect any company that makes a significant commitment to quality to produce 'better' products but this is essentially because we do not understand the difference between quality systems and Quality.

So why is product safety compliance outside of quality systems: and what is the difference between quality systems and Quality? More than 2000 years ago Tao Te Ching wrote '. . . the quality that can be defined is not Absolute Quality' and '. . . Quality and its manifestations are in their nature the same. . .'.

In his book *Lila an Inquiry into Morals,* Robert Pirsig[1] writes: 'Quality doesn't have to be defined. You understand it without definition, ahead of definition. Quality is a direct experience independent of and prior to intellectual abstractions.'

This was echoed by Mr Hiroshi Hamada[2] when he expressed his view that: 'Quality is in the heart, it is in the soul, it is in the mind, and it is in the spirit.'

Personally, I would go one stage further than Mr Hiroshi Hamada and suggest that unless you can feel quality then there will be no training course, or process, on this earth that will help you to achieve it within your company. If we take this concept 'on trust' the 'fact' that we have something that we call Quality that we cannot define, understand or know, and that this Quality 'thing' is made up of lots of other 'bits' that we can define, understand and know. If, for a moment, we accept this concept something rather wonderful happens.

Both ISO 9000 and product safety can be defined: therefore, according to our hypothesis, they must each form a part of Quality. This is important because it allows us to understand that our quality systems and product safety are related – in that they form part of something bigger. That which we constantly attempt to quantify and cannot agree upon. If we put this into an everyday context I can explain two of the many odd 'things' that I have experienced.

1. I was looking through some yachting magazines and found an advertisement for a company making anchors. The advert read something like 'quality anchors fully compliant with BS 5750. . .'.

2. The sales manager of a company told me that his company's product met and complied with *all* the worldwide safety standards because they had '5750'.

What we need to do now is to think what it is that these two examples have in common. Both statements are incorrect: but why? Simplistically, the quality system can exercise its influence only on the process whereas product safety exercises its influence only upon the product. There is a fundamental difference between process space and product space – in that the quality process cannot, by itself, change the physical characteristics of a product. The corollary of this is that product safety cannot directly affect the quality processes within the company. Thus can now explain and understand one of the first causes of confusion about the Low Voltage Directive and CE Marking.

To put this into Rudyard Kipling's words:

I keep six honest serving-men (They taught me all I knew);
Their names are What and Why and When
and How and Where, and Who.

Our company quality plan will show and define the Who, How, Why, Where and When.

It is our product safety plan that will define the What issues, the physical characteristics of our products. And it will do this without any regard for the processes that bring the products into being – if it does attempt to cross this boundary then there is a risk that its purpose will become confused and diluted. However, there is a link between these two. Product safety, being another facet of Quality, will identify parts of its brother, the quality process, what processes may need to change so that all manufactured items comply with the product (safety) standard.

This may all seem like splitting hairs and semantics but these concepts are the very foundations of what we must achieve and I make no apology for labouring them. I shall now explain why they are important – after which it may be useful to reread the above. The vast majority of products[3] that have not been properly assessed for product safety compliance will fail to comply with their appropriate safety standard. If we take our new-found knowledge of the differences between product space and process space then we can conclude that the processes operating within a quality assessed company will ensure that all of its production will be uniformly compliant or – as may be the case – non-compliant.

Checklist for compliant products:

- What harmonized standard is appropriate to your products?
- How many copies are 'personally owned by' your designers?
- Do you have a product safety review checklist?
- Does it mirror the safety standard on a clause-by-clause basis?

- Do the safety test results include abnormal conditions of test?
- Have you defined the safety critical features and aspects of the product?
- Does your quality process verify these facets on product units?

If you do not know the answer to any of these questions then there are two issues for you to consider:

Do my products conform to the Low Voltage Directive?
How can I demonstrate due diligence?

The golden rules

I am always asked: 'You've been doing this for years – there must be some simple tips and methods . . .'

During the early 1990s I tried to produce an 'expert system' to help in product safety analysis. I tried for many weeks to develop a simple structure, after several months I would have been pleased to find any structure that would work – eventually, years later, I gave up. Through the following chapters I shall show some simple techniques to ease design, product safety proving (compliance verification) and to help you conduct physical reviews. But there are no short-cuts or panaceas that replace understanding, knowledge and experience. However, I have learnt a few 'golden rules' over the years (mostly they result from bitter experience), here are a few of them:

1. Always have a copy of the relevant standard with you and use this book to aid your understanding of it.
2. Never review a product, or use a product safety review checklist unless you are confident of your understanding, competence and ability. (You could be placing yourself at considerable personal risk – physically and financially – if you make a mistake.)
3. If in doubt, always err on the side of caution – when standards are revised the changes are usually to make them clearer or the conditions more severe.
4. If you find part of the product is 'marginal' but the interpretation of the standard allows you to accept the design, consider (and explain to your manager) the business implications if that marginal *pass* becomes a marginal *failure*!
5. Remember that critical dimensions are specified as *minimum* or *maximum*. If the dimensions given in the standards are used as design nominals, they will guarantee that 50 per cent of all products will be non-compliant and potentially illegal.
6. Never be pressurized into allowing a dubious product to be supplied to a customer – samples and repaired items must meet the latest standards of safety.

7. Always correct a non-compliance. To place the CE Mark to a product knowing it to be non-compliant is a criminal offence. You may be prosecuted, fined, and imprisoned.
8. Always proceed methodically and record each step. This will allow others to follow your path and will be essential when someone needs to answer a detailed question about the product or to re-specify a critical component.
9. Always get a copy of a safety certificate for each Safety Critical Item before you are committed to using that component.
10. *Never assume.*
11. *Always demand proof.*
12. *Never accept verbal assurances.*

Benefits of using this book

Although this book does not replace electrical safety standards and regulations it offers readers the following benefits:

- Plain English guide to the requirements.
- Easy and detailed guides to suit your production volumes and time-to-market requirements.
- Job function briefings – so that you know what you need to know.
- Years of industry expertise in one handy package.
- Worked examples of creepage and clearance calculations.
- Pointers to Internet and other sources of information.

Health and safety warning

Many of the tests and reviews required by the safety standards described in this book are inherently hazardous. This book is written for the guidance of professional engineers who are expected to have the knowledge and experience to avoid hazardous situations that may arise. Adequate safeguards for personnel and property must be employed when conducting these tests and reviews. The author and publishers provide the following information in good faith and without any liability for any injury or damage resulting from its use or misuse.

Scope

This book is designed to assist manufacturers and importers in ensuring that their data files contain sufficient information to support CE Marking. It covers the detailed technical requirements of one of the major standards in some depth. The standard in question is based on the International Electrotechnical Commission's IEC 950 which covers electrical safety of Information Technology Equipment. While there are other standards covering other types of equipment (see Chapter 6), this book concentrates on one of the most onerous standards with the intention of providing the reader with a thorough understanding of how safety requirements can best be met.

IEC 950 has been amended and adopted by the EU as the harmonized European standard EN 60950. Usefully, it is also the basis for the North American standard (Underwriter's Laboratory Incorporated) UL 1950, and Canadian Standards Association standard CSA 22.2 No. 950.

Limitations of use

The purpose of this book is to assist in the understanding of the LVD and related standards – it is not intended as an alternative to the detailed requirements and of the regulations and standards. Readers are advised to identify the relevant regulations and standards and develop their products with reference directly to these documents. Although this book will prove an invaluable aid in achieving electrical safety effectively, a product which complies with its text may not necessarily be judged to comply with the standards.

The limitations of the guidance notes

In addition to a detailed step-by-step guide to electrical safety, this book contains an easy guide. This provides simple, easy-to-follow steps that will result in a compliant product, at the same time reducing design effort and simplifying safety compliance testing. The strength of these guidance notes is that they reduce development time and design changes, because by following the guidance it will remove many opportunities for designing a non-compliant product. The penalty for using only the easy guide may be that the unit cost is higher than a similar product designed following the detailed guidance notes.

The easy guide is therefore particularly suited to the design of one-off equipment and low volume products. Conversely, the detailed guidance notes will be required for high volume, low margin products. They describe all aspects of the standard in depth allowing the designer to take advantage of low cost design options to minimize the unit cost of the final product. The penalty for using this approach is that greater knowledge of the standard and its interpretation is required and the cost of safety compliance verification can be significantly higher.

Reducing costs and time to market

It is surprising how few people appreciate that most of the whole life costs of a product are fixed early on during the initial design phase. The diagrams below show the relative costs that can be set in stone even before the design reaches the manufacturing department. It is important for us all to be aware of the large amounts of money involved within our company and where it is committed.

With this knowledge, it is a simple process to identify the departments and individuals where any company investment in training or resources have the most beneficial response. Remember that these ratios will depend upon the type of product, the volumes and manufacturing processes. But let us also consider the cost of a simple mistake – it has often been said that 'failing to plan is planning to fail' and so in this book we shall acknowledge that we all are fallible, that we will make mistakes and accept that when they occur they form a significant source of additional costs and delay.

This is an important consideration because if we know that there will be errors – then we know that to look for them is a good and positive thing to do. The cost of not getting it right has not been stated more clearly than during the October 1991 EuroPACE Quality Forum. Mr Hiroshi Hamada, the President of Ricoh, gave the cost of fixing a single defect as:

- $35 during the design phase
- $177 before procurement
- $368 before production
- $17 000 before shipment
- $690 000 on customer site

From this simple example it is obvious that the earlier an error can be identified the more money a company will save.

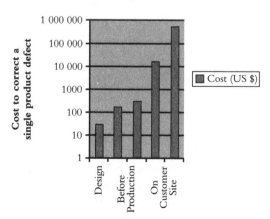

Figure 1.1

It also follows that – since time and money are related – if defects are corrected early then it will take less time for a product to reach production, hence the 'time to market' will be reduced.

Spend versus committed cost

Personally, I believe that the most powerful way to identify where resources (time, money and material) are wasted is to consider the spend during each phase in a project and to compare these to the whole life costs that are committed during that phase. Figure 1.2 shows the classic – textbook – spend curve. From more than twenty years in R&D I 'knew' where the least money was spent and where most savings could have been made on projects that I had been involved in – but I could not prove it from the graph. Several years ago I hit upon the answer.

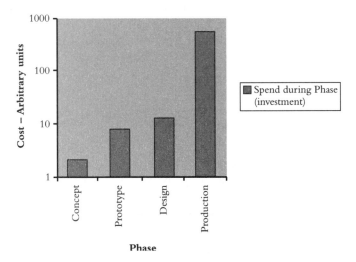

Figure 1.2   *Classic spend during a product lifetime*

We can see that the spend is least during the early (concept to pre-production) stages and that the spend only starts to increase once we hit production. At the end of the product life cycle we can derive a figure for the total cumulative spend. Now comes the interesting bit – we ask our manufacturing engineers how much they were able to change the design of the product to reduce its cost. In the traditional type of industries one would expect a claim of between 5 and 10 per cent. Now we ask the pre-production engineers the same question. Finally we ask the design engineers the same question.

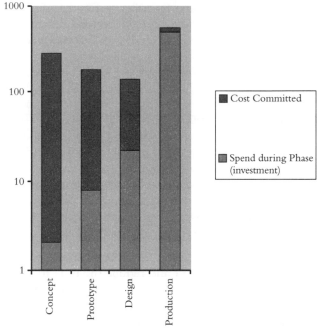

Figure 1.3

We now look at the money invested into the product design at each of those phases and ask – could the total cost have been reduced by spending more during the concept phase. The following are some examples of how money could have been saved:

1. 'I'll design my own power supply to save time.' The time to verify the design of a simple linear power supply can vary upwards from ten days – but plan on taking twenty days. The worst design I saw could not be modified to make it compliant, even after more than fifty man days of engineering effort. Using fully approved power supplies saves design time, and saves compliance testing time; but they will need to be tested in application.

2. 'It was all right in North America therefore it must be OK over here.' Generally if the different frequency or voltage does not cause a non-compliance the use of UL 'Deviations' against the IEC standard will. ('Deviations' are where the North American standard deviates from the requirements of the IEC standard – in many instances these produce a relaxation to the original requirement and the resultant product may fail to comply with the relevant harmonized European standard. The converse is also true. The product designed to our harmonized standards may fail to meet some of the design requirements of North America or Canada.)

3. Creating a critical items list will prevent non-compliant slots and apertures being designed into what may need to become a fire enclosure.

4. Developing the self-discipline to collect safety certificates and reports for critical components during the concept and design phase will save many days of avoidable effort (and avoid additional costs) during compliance verification.
5. Conducting early safety compliance tests and reviews will minimize redesign, scrap and delays.
6. Holding a product safety review to discuss the design concepts will identify and help avoid options that could involve high risks or costs.
7. Holding product safety reviews between project phases will quickly identify potential difficulties and non-compliances and allow quick (cheap) corrective action to be taken.
8. Stopping a project from progressing between phases until safety activities are completed will always reduce risks and may also reduce costs.
9. Balancing compliance-proving cost against production costs may not be easy but it is now an essential task for designers.

Based upon the author's experience, many companies could reduce their design costs by at least 10 per cent and reduce the time to market for most products by four weeks. To achieve these improvements, define clearly the product safety design requirements, ensure that the design team understand the requirement and include product safety review as part of the normal design process.

North America and Canada versus Europe

It is true that the North American UL (Underwriter's Laboratory) Standards are based on the same IEC standards as are our harmonized standards but there are fundamental safety differences between UL listed products and EN certified products.

Some of these are simple and obvious – e.g. the colour or European power cord versus the 'color' of US cordage. Other differences are more subtle and abstruse. Many US transformers are optimized to run at 60 Hz. This is good design practice and minimizes the cost of iron and copper – but they run into saturation and become potential fire and electric shock hazards when operated at 50 Hz. Hence be very careful of all inductive mains wound components designed for 60 Hz operation – e.g. transformers, solenoids, motors etc.

Notes

1. ISBN 0 552 99504 5, first published 1991.
2. Mr Hiroshi Hamada is President of Ricoh and spoke at the 1991 EuroPACE Quality Forum.
3. A major test house has stated publicly that 100 per cent of all new products submitted for testing failed to comply with the required standard.

2 The Directives

The Low Voltage Directive

Electrical product safety has been mandated in Europe since the Low Voltage Directive (LVD) first came into force in 1973 – then, as now, its general requirement can be summarized by the eleventh commandment 'Thou shalt not put unsafe products on the market'. This seems simple but there are many products both in the marketplace and currently in development that will fail to achieve this fundamental requirement. I shall explain this in the following pages.

The 1973 LVD was updated as part of the European Commission's Single Market activities in an attempt to make its implementation more consistent. The original LVD required electrical equipment to comply with minimum levels of safety but because it pre-dated the New Approach Directives there was no mechanism to allow manufacturers (and importers) to signify conformance.

During the early 1990s work began to bring the original 1973 LVD into line with other New Approach Directives so that manufacturers could identify their products with the CE Mark. Since January 1997 it has been mandatory to place the CE Mark on items within the scope of the LVD to signify their compliance.

In January 1995, the LVD was revised, introducing the requirement that the CE Mark is placed on most electrical equipment marketed within the European Union after 1 January 1997. This CE Mark shows that equipment meets the requirements of the LVD and brought the Directive into line with other Directives which harmonize European technical regulations.

Scope of the Low Voltage Directive

The Low Voltage Directive (LVD) applies to items that are rated between 50 and 1000 V a.c. (or between 75 and 1500 V d.c.) and are not:

(a)   Intended for use in an explosive atmosphere, for radiological, medical applications
(b)   Covered by other (international) specialist safety provisions (e.g. ships, aircraft, railway)
(c)   Domestic plugs and sockets
(d)   Electricity supply meters
(e)   (Electric) fence controllers
(f)   Parts for goods and passenger lifts.

What is electrical equipment?

A set of recent (September 1997) guidelines include the words: 'Broadly, it covers consumer and capital goods designed to operate within these voltage limits. These will include domestic electrical appliances, hand-held tools, lighting equipment, electric wiring and installation equipment' – hence the supplier of any non-compliant or hazardous product in these categories may be prosecuted under the LVD. 'However, some types of electrical devices are designed and manufactured for use as basic components of an electrical equipment. Such basic components have no direct useful function for the final consumer' – thus it seems that components will be excluded from the LVD. At first this looks inconsistent as the LVD and other safety directives make it an offence to supply defective and other components that – by their intended use – could cause a hazard to arise. For instance – if I buy a cheap batch of faulty 'X-Capacitors' and supply them directly to consumers or to manufacturers then I could be in breach of this directive. The September 1997 (draft) guidance notes were suggesting that certain non-safety critical components be excluded from this list: 'Integrated Circuits; transistors, diodes and other simple, basic semiconductor devices; passive components such as capacitors, inductors, resistors, and filters; connectors; and flexible conduits.' Some electromechanical components are also suggested: relays and microswitches. We need to note that:

1.   While this list of components suggested for exclusion from the LVD includes potential Safety Critical Items – opto-isolators, connectors, relays and filters – the harmonized standards will usually contain a specific requirement that Safety Critical Items must comply with the relevant IEC component standard.
2.   The wording of the draft proposal continues: 'Moreover the safety of the final user or consumer will depend on how the components are integrated into the final product, their use in application and the overall characteristics of the final product. Therefore the LVD does not apply

to these components as such and the manufacturer of the final product must accept liability with regard to the provisions acceptable to his product, on the basis of the principles governing the placing of the equipment on the market for the purpose of distribution or transference to the final user or consumer.'

3. Any product or component that is not within the scope of any of the New Approach Directives – such as the LVD – is automatically within the scope of the General Safety Directive. It should come as little surprise to us that the General Safety Directive makes it a criminal offence to place an unsafe product on the market.

4. 'However the scope of this exclusion must not be extended to items like lamps, starters, fuses, etc., which, even if they are often used in conjunction with other electrical equipment and have to be installed in order to deliver a useful function, may be directly or indirectly intended (and directly relevant in terms of safety) for the final user or consumer and are themselves to be considered electrical equipment in the sense of the Directive.' This double negative is likely to cause as much confusion for readers for whom English is not their native language as it does to those for whom it is! What I think the suggested draft amendments are trying to express is that 'items like lamps, starters, fuses, etc.' should be retained within the scope of the LVD.

5. Finally, 'In addition, an intermediate group of items exists which for reasons of their possible destination or for needs of continuity in the traditional regulatory regime complies to them under this Directive since 1973, have to be considered to be covered by the LVD. These are items like cable management systems, switch gear, connection boxes, ballasts, etc.'

We must remember that the range of products covered by the LVD is awesome and it is not easy for the LVD to express its scope more clearly than this – however, as a rule of thumb it may help to remember:

To Mark	or	Not to Mark
It is a criminal offence not to mark a product covered by the LVD		It is a criminal offence to mark a product that does not comply with the LVD
You may place the CE Mark on any product that complies with any RELEVANT New Approach Directive		You may be required to remove the CE Mark from a product not covered by any New Approach Directive

If this treatment appears slightly perverse then please accept my apology – the reason being to take an extreme approach from which to view the options. By taking the situation outside of the 'grey' areas it is often easier to gain a clearer understanding of what is intended by the legislation.

New Approach European Directives

The CE Mark signifies that the product displaying it meets all relevant European Directives. The harmonization of regulations means that a product can go through a single development and test regime, and can then be sold throughout the European Union without having to meet further requirements. Other Directives with which a product may need to comply are shown in the following table.

Directive	Directive number	Mandatory date for CE Mark
Toys	88/387/EEC	Jan 1990
Construction products	89/106/EEC 94/23/EC	Jun 1991
Simple pressure vessels	87/404/EEC 90/488/EEC	Jul 1992
Telecomms terminal equipment	91/263/EEC	Nov 1992
Machinery	89/392/EEC 91/368/EEC 93/44/EC	Jan 1995
Active implant medical devices	90/385/EEC	Jan 1995
Satellite earth stations	93/97/EEC	May 1995
Personal protective equipment	89/686/EEC 93/95/EEC	Jul 1995
EMC	89/336/EEC 92/31/EEC	Jan 1996
Gas appliances	90/396/EEC	Jan 1996
LVD	73/23/EEC	Jan 1997
Hot water boiler efficiency	92/42/EEC	Jan 1998
Medical devices	93/42/EEC	Jun 1998
Recreational craft (boats)	94/25/EC	Jul 1998
Proposal marine equipment	95/C218/06 96/C101/07	(Jan 1999)
Lifts	95/16/EC	(Jul 1999)
Proposal pressure equipment	94/C246/01 94/C52/04 94/C207/05	(Jul 1999)
Proposal in-vitro diagnostic medical devices	95/C172/02	(Jul 2002)
Non-automatic weighing machines	90/384/EEC	Jan 2003
Explosives for civil use	93/15/EEC	Jan 2003
Safety equipment for use in potentially explosive atmospheres	94/9/EEC	Jan 2003
Proposal cableway installations for passengers	94/C70/07 96/C22/06	
Proposal efficiency for household refrigerators and freezers	96/C120/02	

It is important to realize that the manufacturer has the responsibility of determining which directives are applicable. It should be noted that in many cases where the Low Voltage Directive is cited on the Declaration of Conformity that the EMC Directive should also be listed – and vice versa.

Unlike previous safety legislation, the LVD imposes formidable penalties upon businesses and individuals that do not comply. The greatest penalty may be the financial impact of having the sale of products prohibited in the whole of the European Union; however, there are other, more direct, penalties which include:

- enforced recall and corrective action
- individual prosecutions under criminal law
- heavy personal fines
- heavy corporate fines

These are the risks faced by companies and individuals that ignore this important legislation. Hence it is vital that manufacturers instruct and train their engineers to understand and implement compliance engineering.

Who are manufacturers?

The words of the Directives vary slightly but the producer is generally taken to mean:

> The manufacturer of the product, when he is established in the community, and any other person presenting themself as the manufacturer by affixing to the product his name, trade mark or other distinctive mark, or the person who reconditions the product;
>
> The manufacturer's representative, when the manufacturer is not established in the community or, if there is no representative established in the community, the importer of the product;
>
> Other professionals in the supply chain, insofar as their activities may affect the safety properties of a product placed on the market.

We can usually consider that producers, manufacturers, importers, wholesalers, retailers, mail order traders, auctioneers, equipment repairers and second-hand suppliers are covered by the LVD when they deal with the supply of electrical equipment.

However . . .

The comments within this section are not intended to be alarmist but I merely report them as a simple statement of fact. The observations are important to us all as designers, as OEMs, as importers or as exporters.

It has been estimated that many of the products that have not been reviewed by an independent product safety agency (such as TÜV or BSI), and are currently in manufacture, will fail to comply with the safety standards that are relevant to them. One well-known test house reported that 100 per cent of all products that it received for safety testing failed – I would endorse those comments as my own personal observations support these findings.

Having reported such a significant finding I believe that it is reasonable to make the personal observation that a significant number of these non-compliances will escape the description of 'life-threatening': but there are a finite number of products whose non-compliances could cause injury or damage to property.

So how can this situation have come about?

Here we can only speculate on possible causes:

Awareness: Everybody knows that the new product must conform to the latest standards but nobody writes that need into the design specification.

Lack of resources: To meet the project goals it is necessary to update a previous design and there is only time to consider the functionality.

Training: The company cannot afford to divert its engineers from the task in hand – or does not have the money – or fears that training makes its engineers more marketable and that it will lose them to competitors.

Lack of reinforcement: The LVD was implemented into EU law in 1974 – the fact that there is so little training available or history of enforcement action is sufficient evidence to make further comment unnecessary.

So if there are that many non-compliant products, why aren't there more accidents?

A reasonable question and the answer is just as reasonable. As designers we take pride in our work and we are aware of obvious safety hazards – such as exposed hazardous voltages. So as professionals we avoid design features that would be an immediate and obvious hazard.

However, life is not as simple as that! So, let us analyse a simple non-compliance that will occur during the development phase of a new product. This analysis is detailed and illustrates the depth of knowledge, experience and attention to detail that is necessary when investigating any product safety of a design.

Example: A wire carries low level signals between a circuit board and an *operator accessible* connector. The signals are derived from circuitry powered by Safety Extra Low Voltage (SELV – less than 60 V d.c.) and the power levels are less than the level permitted by the relevant standard (a typical level is 15 W *maximum*). Thus we classify the circuits as Safety Extra Low Voltage Energy Limited – or SELVEL. This classification is important because the operator may touch circuits and parts carrying SELVEL.

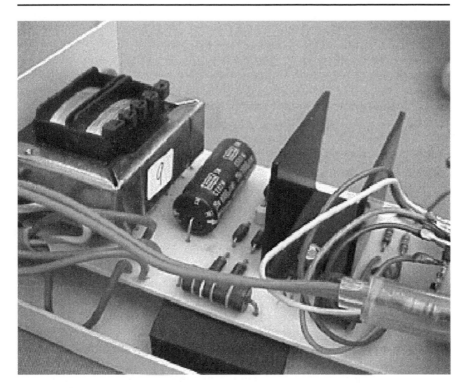

Figure 2.1

We now consider a single failure – the wire breaking. The break will usually occur where the solder has 'wicked' into the strands of wire, combined them into a single mass and weakened them. The wire breaking is a reasonably foreseeable[1] single point failure. 'So a wire breaks – Well big deal! So it stops working!' If the wire is 'broken' we *must* consider what it 'could' touch. This is important so I will say it again '. . . consider what it *could* touch . . .' Personal advice: Before carrying out a safety review do what I do and say to yourself: 'If I make a mistake, or miss something, someone may die!'

The correct attitude for a safety review is to be *critical*. This does not mean negative and destructive: being critical means that we must identify *all possible* non-compliances and hazards. This will mean 'unlearning' all the positive attitudes that you have learnt and using your education, experience and intellect to consider all the bad things that can expose the operator (and service personnel) to hazards. Only when all the potential problems are identified and recorded can we give ourselves permission to consider their probability and their solutions. *Never* attempt carrying out any other analysis during a safety review.

Developing a *critical* 'eye' is the most important step to product safety – we must learn to see the product through another set of eyes. It is not an attempt to consider the probability or the likelihood of any particular contact: it is our responsibility to ensure that the loose wire cannot physically come into contact with any hazardous source.

To assess this we move the free end of the wire in *any* direction. The purpose of this is to determine if it is possible to contact hazardous voltage or hazardous energy – or to come within a few millimetres of these hazards (i.e. within permitted creepage and clearance distances). If it can be moved to contact a hazard then the operator accessible connections may – under a single failure condition – become hazardous and could become lethal.

So why are there so few 'accidents'?

From our recent example many readers will be thinking 'this guy cannot be serious . . .' or even 'if this is true then why aren't we knee deep in bodies?' Personally I don't blame them – those were the same thoughts that I had when I began working in product safety after twenty years spent in R&D. But wait. Let's try to put some numbers into this scenario.

The failure rate for these events will be less than one failure every million hours (1 fpmh to the purist). So the probability of this one product becoming hazardous is *very* low. So now let us try to estimate how many million pieces of electrical equipment come into service each year? How many hours will they operate during their ten year[2] life?

Without going into a formal statistical analysis, if we take a small number and multiply it by two *very* large numbers then it is obvious that there is a finite – and therefore a real – probability that someone will become the victim of an 'accident'. It is these statistics that explain why it may take many years, before a non-compliance will manifest itself as a hazard, and why not all non-compliances have fatal consequences.

The incentive for product safety

To use the same language as the masters of personal fitness: 'No pain: no gain!'

No pain: no gain – the past

It is another sad truth that if 'laws' are not enforced then the system breaks down and the 'laws' are ignored. How many cars pass us on the motorway when we are travelling at the maximum legal speed? What incentive was there for engineering courses to include product safety? What incentive was there for companies to commit precious resources to product safety training and implementation. It is my personal belief that the lack of consistency in worldwide enforcement is the reason why many companies failed to take the 1973 LVD seriously.

No pain: no gain – the present

Since the introduction of the EMC and Low Voltage Directives we have seen significantly more activity in inspecting, reviewing and testing products that cross our European borders. The result has been an upsurge in training and regulatory compliance activities within our industry. Consider the personal thoughts and deliberations of a company executive who discovers a major safety defect in a high volume product that he has been shipping for ten years.

No pain: no gain – the future

I am sorry to keep going back to reliability theory and failure rate but this will be – I promise – the last time I mention it . . ..

As more electrical products become commonplace in our homes and businesses then the probability of more accidents 'happening' will also increase. As most of these electronic and electrician products have operational lives of ten years then the build-up of products is increasing at a seriously high rate.

Example: When I started in electronics design (in the 1970s) I had a scope, a power supply and a soldering iron. Our laboratory did not have a photocopier and there was one electronic calculator (using reverse Polish notation) between forty engineers. Typewriters were manual, the company computer sat at the other end of a telephone line and the keyboard was a teletype terminal. No I'm not looking for a sympathy vote – I'm trying to put the present situation into perspective by showing the enormous amount of everyday electrical equipment that has built up in twenty something years . . ..

There is a tremendous growth in the amount of electrical equipment coming into the world and to keep the level of risk low for the user (that's us) then the safety standards of twenty years ago would not be good enough. This is why these safety requirements have got tougher and why they will continue to get tougher. Thus the trend towards tougher standards is, in the author's humble opinion, likely to continue into the foreseeable future.

The CE Mark

The form of the CE Mark is shown right. Only the letters are used (the grid is for reference). The proportions and spacing must be maintained and the mark must be at least 5 mm high.

Figure 2.2

The CE Mark and demonstrating compliance

Most of the New Approach Directives allow manufacturers and importers the option of self-declaration. Here the terminology and implementation differs slightly, dependent upon whether the product is manufactured inside or outside of the EU – so please refer to the section and the Declaration of Conformity that applies to your particular circumstances as an EU manufacture or non-EU manufacture.

Non-EU Manufacture

Example of a typical Declaration of Conformity, for a Non-EU Resident Manufacturer to the Low Voltage and EMC Directives.

COMPANY LETTERHEAD

NAME AND FULL ADDRESS

Manufacturer's Declaration of Conformity with:

We declare, under our sole responsibility, that the products identified in this declaration, and to which this declaration relates, are in conformity with the requirements of Council Directive:

   sample text 73/23/EEC as amended by 93/68/EEC on the harmonization of the laws of the Member States relating to electrical equipment designed for use within certain voltage limits.
   Mark affixed January 1st 1997

   sample text 89/336/EEC as amended by 92/31/EEC on the approximation of the laws of the Member States relating to Electromagnetic Compatibility
   Mark affixed January 1st 1996

Description of Equipment:

Product description (e.g. Electric Eraser, Computer Monitor, Power Supply, Stapler, Modem, Document shredder)
Model Number   (e.g. PBX 316-H)
Rating   (*sample text* 230 V 50 Hz 65W) or (*sample text* 230 V 50 Hz 5 A)

Batch of product covered:

SERIAL NUMBERS-   From: *5,000*   To: *7,000*
(Explanatory note: You are advised to put some controls in place that will prompt the *questions "What changes have occurred that may affect Safety or EMC?" "Do we need to re-test or re-examine the product?"* Placing some limitation on the life of the Declaration of Conformity is a very effective method of achieving this goal without increasing the complexity or the existing quality system.)

Standards applied:

sample text Safety   EN 60950 issue 2 Amendment 3, including National deviations for Denmark
sample text EMC   EN 50081-1 Emissions
sample text   EN 50082-1 Immunity

The Authorised Signatory to this declaration, on behalf of the manufacturer, is identified below:

Name   A C Smith
Title VP Engineering
ACME Computer Systems Inc.,
Address P.O. Box 12345, Rockfield, IL 94812 USA

Signature   *AC Smith*   Date

The Responsible Person, based within the EC, is identified below:

Name
Title
Address
Signature   Date
(Explanatory note: The Responsible Person MUST BE a resident of the E.U.)

Figure 2.3   *Declaration of Conformity for product of non-European Union manufacture*

EU Manufacture

Example of a typical Declaration of Conformity, for an EU Resident Manufacturer to the Low Voltage and EMC Directives.

COMPANY LETTERHEAD

NAME AND FULL ADDRESS

Manufacturer's Declaration of Conformity with:

We declare, under our sole responsibility, that the products identified in this declaration, and to which this declaration relates, are in conformity with the requirements of Council Directive:

> *sample text 73/23/EEC as amended by 93/68/EEC on the harmonization of the laws of the Member States relating to electrical equipment designed for use within certain voltage limits.*
> *Mark affixed January 1st 1997*

> *sample text 89/336/EEC as amended by 92/31/EEC on the approximation of the laws of the Member States relating to Electromagnetic Compatibility*
> *Mark affixed January 1st 1996*

Description of Equipment:

Product description (e.g. Electric Eraser, Computer Monitor, Power Supply, Stapler, Modem, Document shredder)
Model Number   (e.g. PBX 316-H)
Rating   (*sample text* 230 V 50 Hz 65W) or (*sample text* 230 V 50 Hz 5 A)

Batch of product covered:

SERIAL NUMBERS   From: *5,000*   To: *7,000*
(Explanatory note: You are advised to put some controls in place that will prompt the *questions "What changes have occurred that may affect Safety or EMC?" "Do we need to re-test or re-examine the product?"* Placing some limitation on the life of the Declaration of Conformity is a very effective method of achieving this goal without increasing the complexity or the existing quality system.)

Standards applied:

sample text Safety   EN 60950 issue 2 Amendment 3, including National deviations for Denmark
sample text EMC   EN 50081-1 Emissions
sample text   EN 50082-1 Immunity

The Authorised Signatory to this declaration, on behalf of the manufacturer, is identified below:

Name A C Smith
Title Technical Director
ACME Computer Systems Inc.,
Address P.O. Box 123, South Field, North Midlands, RD4 15LX England

Signature   *AC Smith*   Date

Figure 2.4 *Declaration of Conformity for product of a European Union manufacture*

When reading the following sections it is important to appreciate that the Low Voltage Directive is intended to provide only general and basic guidance. The exact implementation of these requirements may also vary on a country basis and are most certain to change as the result of case law involving 'problems' and legal actions.

There are several ways to go about satisfying the EU requirement but the subject that seems to have caused much confusion is 'Who needs to sign the Declaration of Conformity?'

Perhaps the easiest way to explain this is to consider what the Low Voltage Directive is attempting to achieve. To ensure that only safe products are placed on the European market three things are essential.

1. We must ensure that the product is designed to be safe.
2. We must ensure that all future products will also be safe.

3. A named person must take full legal responsibility and full personal liability, under European Law, for the product.

To meet the first two requirements we must nominate someone within the company who has technical knowledge and responsibility for the product, and the authority to declare compliance on behalf of the company. This person will sign the Declaration of Conformity as the 'Authorized Signatory' on behalf of the company and will be expected to have been given the appropriate technical control and authority to maintain the appropriate level of safety compliance.

They may work, live and reside anywhere in the world: they do not need to be a resident of the EU. (Typically this person could be the CEO, the VP of Engineering, the Quality Director. There are good reasons — explained below — for placing this responsibility outside of 'manufacturing' but the LVD would seem to accept a signatory from within the manufacturing area.)

To meet the third requirement — a 'Responsible Person' taking personal liability under EU law — it is obvious that the person must reside within the EU. The reason for this is simple — if they reside outside the EU they cannot be made the subject of European law. Hence the 'Responsible Person' must be a resident of the EU.

It is generally excepted that the Responsible Person will have little influence over the 'foreign' manufactures, but they will be expected to ensure that the Declaration of Conformity is valid and that a copy is available for inspection by the Enforcement Agencies. Clearly if the Responsible Person fails to produce a valid Declaration of Conformity — or there is evidence that they placed a product on the market in the knowledge that it did not comply with the general safety requirements of the LVD — then there may be a case to answer.

It should be noted that whereas there may be no legal requirement for the Responsible Person to hold a copy of the Data File there are several strong reasons for the Responsible Person holding a copy of the Data File and the Declaration of Conformity. The Declaration of Conformity is a legal document and signing it must not be taken lightly. It is a statement of technical compliance that says the product meets harmonized safety requirements and can therefore be sold anywhere in Europe. By applying the CE Mark, the manufacturer makes a commitment to supply proof of compliance to the authorities. This evidence is referred to as the Data File. The signatory of a Manufacturer's Declaration of Conformity is personally liable in criminal law if an offence is committed (fraudulent use of self-declaration, or placing a non-compliant product on the European market are both criminal offences).

Compliance

The LVD defines only general requirements. It leaves the detail of how to achieve compliance with these requirements to the manufacturer and the signatory(ies) of the Declaration of Conformity. There are several routes to the CE Mark but essentially there are two potential and widely different approaches. These we shall refer to as the Standards Approach and the general declaration approach. The Standards Approach uses a relevant Harmonized Standard and demonstrates conformance on a clause by clause basis to that standard. This is the approach favoured by the test houses and product safety agencies. The advantages with this approach is that the methodology is the most common – hence it is the most widely 'accepted' method and automatically provides a significant amount of information to include in the Data File.

If we read the directive carefully we can infer that it is possible to complete a Declaration of Conformity stating that the product meets the 'general requirements' of the Directive and not to refer to any standard. This would, however, give us substantial problems if called upon to produce a Data File – remembering that we may face criminal prosecution if the enforcement officers are not satisfied with our technical explanations, and our documentation provides evidence that we have not observed 'due diligence'. I am not aware of anyone taking this path and I would personally not give the option a moment's thought, other than to dismiss it instantly in favour of the standards route.

Compliance and complacency

We must note carefully the wording of the LVD in the use of standards. By demonstrating compliance with relevant harmonized standards we are 'deemed' to comply with this and other safety directives. The use of the word, where it applies, is important because it does not guarantee us a defence from prosecution. If we design a product, test it to a relevant standard and someone is injured then we may still be prosecuted under civil law – for damages – and criminal law – for manslaughter or negligence. This is particularly relevant in two cases:

1. Where a product is state of the art and standards do not exist.
2. Where a product crosses several technical barriers or interfaces.

Point 1 is outside the scope of this book but point 2 occurs surprisingly frequently.

Here is another example. Let me play devil's advocate and consider the situation from how I might attack the compliance strategy of a large, three-phase product that is computer controlled.

1. The safety of computer operators is assured by EN 60950, but the computer may be directly connected to a piece of equipment that has not been evaluated to EN 60950.
2. The operator receives an electric shock while changing the printer ink cartridge and was killed.

As expert witness for the prosecution I would point out:

1. That the operator had accepted a level of occupational risk by choosing to use the computer.
2. The level was defined by EN 60950.

I would examine the electrical interface within the 'equipment' and attempt to show that this design of the 'equipment' did not provide the same level of safety as EN 60950. This leaves the possible argument:

1. The digital output was intended to be used with information technology and office equipment.
2. The relevant Harmonized Standard for the interface is EN 60950.
3. The designer failed to take note of this.
4. The designer had a duty of care not to place the operator at unnecessary risk.
5. The designer was negligent and failed in that duty.
6. The manufacturer placed a non-compliant product upon the market.

The reason for this laboured approach is to explain some of the considerations that we designers are now being faced with in our daily work. The solution in this example was to assess the interface (or the entire product) to EN 60950. Record our design decisions and ensure that they are preserved for at least ten years after the last product is sold.

Choosing a standard

This is not a simple task. At the time this book was written (1997) there are more than 630 Standards and Amendments formally listed in the *Official Journal of the European Communities* (affectionately known as the OJ). Choosing the appropriate one is not an easy task but help can be obtained from most test houses, BSI Technical Help for Exporters or from www.gkcl.com.

Please check that you have read the previous section and understand the potential overlap between standards.

Placing on the market

'Placing on the market' means making it available. In other words an item for sale – offered as a free sample – or on a sale or return basis, or for hire, or after must be safe and must be CE Marked.

Second-hand goods

All second-hand goods are covered by the requirements of Article 2: '. . . that electrical equipment may only be placed on the market if . . . having been constructed in accordance with good engineering practice . . . it does not endanger . . .'.

The implications here are significant since the requirements of current harmonized standards may greatly exceed those of the original national standards to which the product should have been designed. Hence, any person selling (or auctioning) second-hand goods must be aware of their legal responsibility to ensure that the goods are safe within the *current* meaning of the LVD and its harmonized standards. As there is little case in law it is not possible to provide detailed advice – and therefore the comments in this section must be treated with *great caution*.

It is possible that the use of warning notices and labels[3] may provide little or no defence. It may be more useful to give an example of my reasoning and analysis than merely to give details that may (or may not) prove to be correct following some future court action.

One way we can evaluate the situation is to consider the technical argument that we might use if we were prosecuting the case:

1. The warning was on a removable label – once removed the warning was gone.
2. The warning couldn't be understood by children or the illiterate.
3. The unit was supplied with a normal mains cord and plug, and was presented as a functioning, working unit.
4. The fuse was removed – but could be replaced by the user.

So if I really want to supply this product in a non-working unit how would I do it? I think I would start by finding something that 'looks' like a mains cable but *did not* contain *any* copper conductors. This would make it impossible to get hazardous voltage anywhere near the user.

We can have a moulded plug fitted to one end and even glue the fuse holder in place – because the plug could be connected to the mains supply it is important to use a 'real' one that will not cause a hazard, for example by leaving broken bits in the mains socket. The equipment end of the 'cord' would be terminated inside the unit and all traces of the internal wiring would be removed. Hence the user would have no reasonable way of connecting the unit to the mains supply.

Once again, this is an example of the sort of analysis that we must conduct *and document* as evidence. We must ask 'How can someone reasonably make the product hazardous, how might I expect the equipment to fail and what can I reasonably do to avoid a hazard occurring?'

A label will warn the purchaser that it is a non-functioning unit. But we cannot consider this as a safety warning because it might become removed. Moreover we must be cautious of the following wording found within the General Safety Directive in Article 3 'Provision of such warnings does not, however, exempt any person from compliance with the other requirements lain down in this directive.' True it is not from the LVD but if I was acting as expert witness for the prosecution that is one route that I would suggest.

- Permanently fixing the fuse holder will prevent the user fitting a fuse.
- Using non-conductive wire will prevent the purchaser from cutting the dummy plug off and fitting a real one.
- Removing internal wiring will prevent anyone from transforming a potentially non-compliant equipment into a potentially lethal product.

This analysis may be more than is strictly necessary within the current scope of the LVD – but it illustrates the considerations that we should make and document.

Repaired goods

According to the LVD it would have been illegal to send defective goods for repair, but there is an exemption to companies and individuals sending defective products for repair. *We must attach a warning* to the repaired item stating that it is defective and may be hazardous. Since this warning will be read by service personnel we can be reasonably certain that it will be responded to correctly.

The repairer is not required to fit any CE Mark on the product but it is important to remember that the user may rightfully expect that the newly repaired equipment will comply with the latest harmonized safety standards. This is unlike current EMC legislation – here we ensure that the degree of risk that the operator would expect, if using new equipment, is no greater if the equipment has been repaired or refurbished.

It is also useful to note that the Unfair Contracts Act will prevent a contractual agreement to supply or to accept equipment to an inferior level of safety.

Notes

1. These very words will be found in many of the New Approach and old-style European Safety Directives.
2. An individual may sue – under the Product Liability Directive – for an injury caused by a defective product for ten years after the date of purchase.
3. E.g. 'Do not connect the mains – for decoration only'.

3 Briefing notes to key job functions

Director's brief

Someone within your company will be signing a Manufacturer's Declaration of Conformity or Incorporation and, in law, is known as the Responsible Person. This individual accepts *personal liability* for the products that they authorize for CE Marking, and they can be pursued in criminal law. The Responsible Person must have sufficient authority, knowledge and training to carry out their tasks adequately. Consider a scenario where the Responsible Person has incorrectly authorized CE Marking, and the person using the equipment has been injured. If the Responsible Person is sued under criminal and civil law, he could face three months' imprisonment, a £5000 fine, costs of around £25 000 and damages of £50 000.

If you do not protect this key employee they could be imprisoned, lose their house and become bankrupt. The usual safeguards against this scenario are:

Training – if the Responsible Person understands their responsibilities and how to carry them out then they will be better placed to protect not just themselves, but your company as well.

Authority and confidence – the Responsible Person must have the last word on shipping hardware. This may cause friction between the Responsible Person and the manufacturing department. It is vital, for your business, that the Responsible Person has the confidence that you will back their judgement.

Professional Indemnity Insurance (PII) – this may be useful to support a legal defence and claims for damages but it is *not* an answer in itself. The reasons for this are twofold. If you have a history of claims then you will become uninsurable and, in any event, PII will not prevent you from going to prison.

The Responsible Person is a key guardian of your business and needs to be a high calibre individual who is not swayed by pressure.

Responsible Person's brief

Congratulations – you have one of the most responsible and interesting jobs in your organization. That is why your brief is placed high on the list. You may also find yourself under enormous pressure to concede items that are almost, but not quite, compliant. When this happens, my personal advice is to do as I do and say to yourself: 'I can go to prison if this kills someone.'

It is vital that, within your organization, you have line responsibility to 'god'. This may be the managing director or the chief executive, but you should *never* report to a person with responsibility for manufacturing. You need to understand clearly the reason for this. If you find a shipment of non-compliant products then you must be able – for the company's and your own sake – to stop that shipment.

I have held a multi-million pound shipment for safety reasons and if you ever have to do the same, believe me, your actions will soon be noticed. You may find your popularity waning fast in some quarters, and receiving hearty support from others. If you are right, you must not yield to the natural pressure to concede the shipment (because of your legal responsibilities under criminal law). You also do not want to jeopardize your career. For these reasons it is important that (in this rare event) you report at the most senior (non-manufacturing) position practicable, and have confidence that this person will support your technical judgement. If any one of these conditions is not met then you must seriously consider if you should accept the liability imposed by this task.

On a brighter note, this job can be one of the most interesting and challenging within the company, and if you (occasionally) perform your own audits (highly recommended) you will find yourself crossing the boundaries between quality, engineering, purchasing and manufacturing.

The questions you ask will be highly focused, and because of that you will often find weaknesses in procedures. This can make you an even more valuable person, but remember that no matter what you find, your primary purpose is to protect your customers (yourself and your family) by ensuring that only safe products are shipped.

Brief for manufacturing personnel

The new role of Responsible Person may cause ripples of concern for you and your staff. It is natural for you to feel anxious about the Responsible Person checking into areas that, historically, were yours alone. But you should not feel threatened. With your help and support the Responsible Person will become your greatest ally. Consider the following questions:

- Who carries the legal responsibility if a product is non-compliant?
- If you had that responsibility, would you want to verify compliance yourself?

- Would you be afraid of upsetting the person whose work you were checking?

These are the issues that will motivate the Responsible Person and they must be given the right to fulfil their duties. If the safety of a product is affected by changes in design or the manufacturing process, then they must be corrected and shipments may be suspended until the product is compliant. Most important of all – any production lost is not the fault of the Responsible Person: they are merely identifying a potential problem.

Brief for the quality function

Your working relationship with the Responsible Person is vital although you may also feel this to be a threat to your traditional job function. However, the Responsible Person will bring an important gift to quality. Whereas quality has always offered intangible benefits, the CE Mark demonstrates that in the area of electrical safety, at least, your products can be sold within Europe. By working with the Responsible Person, quality staff can identify weaknesses and errors before they become problems which will halt production and demonstrate a tangible added value to the business operation. Areas to check are:

- that the design process is followed
- that safety reviews are held
- that actions are completed
- that purchasing and manufacturing processes are followed
- that change control is watched like a hawk.

Ensure that you only monitor and verify processes. Do not attempt to become involved in the detail of design or compliance verification.

Engineering manager's brief

It is your job to provide the Responsible Person with information for the Data File required to demonstrate conformity with the LVD. Such data should fulfil two objectives. It must provide evidence that the product complies with the requirements of the relevant safety standard and the essential requirements of the LVD. It must also define all safety critical aspects of the product and all safety critical components used within it. As with all European Directives, such information may be requested by the enforcement authorities and in the event of an accident involving one of your products, it will may be used to verify that all appropriate safety measures were taken.

The data you provide the enforcement authorities may also be checked against the products actually being manufactured. This means that changes which affect the safety of the product will require assessment and additional testing and design information may need to be added to the Data File.

It is not a legal requirement to have assembled the Data File at the time of product development – only that it can be made available to the enforcement authorities, should they request it. If you produce consistently safe goods, you may never be asked to present this documentation, but there are other valid reasons for completing it. If you complete this verification exercise before manufacturing begins, you can test the documentation and it will start to save time and money in design and manufacture. Here are some reasons why you should prepare a Data File at the time of product development:

- Thorough documentation will provide the best defence in legal proceedings.
- A mistake that makes the product non-compliant will cost time and money to put right – the later it is noticed, the more it will cost. Properly documenting the design is a good way to identify and correct such errors before production.
- A non-compliance which goes unnoticed because the documentation was not completed may cause great expense in terms of product recall, rectification and damage to the company's market image.
- Production staff can use the descriptive report (I like the term 'Product Description') within the Data File to verify their own work (a vital step in implementing total quality management).
- It is a cost-effective way for inspectors to check safety compliance.

Another important part of the engineering manager's job is to ensure that staff understand the goals in safety compliance. It is surprising that many engineers are never told what safety standards their design must meet. Few have studied the requirements of EN 60950 or other safety standards, yet the phrase 'get it right first time' is common across the industry. The most important job for the engineering manager is to provide an environment for improvement – otherwise you may spend a lot of money on an extensive training programme when what is needed is a little and often training strategy.

Product Safety Compliance is more important than pure functionality. Failure to function just 'disappoints' customers: failure to meet basic safety requirements may 'kill' them.

Designer's brief

In the good old days, the designer only had functionality to consider. Then came styling, aesthetics, reliability, total quality management. Now, you are facing a new batch of regulatory requirements covering safety, electro-magnetic compatibility and, for some products, telecomms regulations.

Complying with these requirements is mandatory and designers with specialist knowledge in these areas will become invaluable. Understanding EN 60950 and other safety standards is not difficult but, because of the mass of information, it is not easy either. Review and assessment of a product to a safety standard requires a detached perspective. No longer are you willing your creation to function correctly – you must learn to use your expert knowledge to devise cunning ways of making it fail. You will be devising ways that an ordinary operator could cause a hazardous situation to arise. If a tool to unfasten parts of the product is provided to the user, then your role play must include unfastening all those parts. You should also ignore warning labels in any language which will not be understood in the countries in which the product will be sold. The key to assessment is to check safety for reasonable misuse and foreseeable abuse.

When reviewing designs we *must* consider what the operator (and service engineer) might do to create a hazard. Can they connect plugs incorrectly, put papers on the top and block ventilation slots, set it to the incorrect operating voltage etc.

4 Directives, standards and essential requirements

The basics

The following general requirement features in many of the harmonized standards and so they make a good starting point to help us to understand what the LVD is attempting to achieve.

Operator

The term 'operator' usually applies to any person who may reasonably come into contact with the product. In the office this may include cleaners and other non-technical staff. In the home it may include children. In the simple world of product safety there are two types of people, operators and service personnel. It is imperative that our product handbooks and documentation acknowledge this fact and do not expect the operator to conduct service type actions.

Two levels of protection

The operator is not usually expected to have any knowledge of electrical hazard and must be given two levels of protection (e.g. basic insulation plus supplementary insulation; basic insulation plus safety earth ground). The reason for these is to ensure that no single fault will create a hazard for the operator.

May contact SELVEL

The only voltages that the operator may contact are separated from hazardous voltage (e.g. mains) by two levels of protection and must not themselves constitute a hazard (e.g. safety extra low voltage and energy limited – typically this will be a source of less than 60 V d.c., less than 15 W, less than 20 J and less than 8 A). The purpose of these restrictions is to make sure that any external accessible part (e.g. the pins of a D-type connector) cannot present a hazard if touched by the operator.

Service personnel

Service personnel are defined by several standards as 'Persons having appropriate technical training and experience necessary to be aware of hazards to which they are exposed in performing a task and of measures to minimize the danger to themselves or other persons.' These words can usefully be employed in our service documentation to ensure that our customer uses only people with the correct training to perform service tasks.

New Approach Directives

There are a number of New Approach (CE Marking) Directives covering almost every type of product – the table on page 14 shows the complete list of Directives which apply to most products at the time of writing.

One facet of New Approach Directives is that they only provide a very top-level description of the safety requirement. They outline, usually in a section entitled the 'essential requirements', and define various means by which conformity with them can be assessed. The European Commission mandates the creation of harmonized standards to support each Directive. These define the technical requirements in more detail. Complying with the standards is usually the simplest way of demonstrating conformity.

'Old-style' Directives

While there are a few New Approach Directives there are many hundreds of directives covering topics such as the 'administrative provisions relating to direct insurance', the 'pollution by emissions from motor vehicles' and most significant to us all 'general safety'. The General Safety Directive is so important to us in engineering that its implications are dealt with in a following chapter.

Harmonized standards

European harmonized standards are created by CEN (Comité Européen de Normalization), CENELEC (Comité Européen de Normalization Electrotechnique) whose standards cover electrical devices, and ETSI (European Telecoms Standards Institution) whose standards cover telecommunications products. Wherever possible, these organizations will adopt an existing international standard to be used as a European specification. For instance,

EN 60950 is the harmonized European version of the International Electro-technical Commission's IEC 905.

Once a standard has been created by the European standards bodies, the European Commission will adopt it (assuming it is happy with the detailed content of the specification) and publish its reference in the *Official Journal of the European Communities* (OJ). It may then be used to demonstrate conformity with the Directive against which it has been adopted. This is a very simplified view of the standards making process: in practice there is much interaction between national, European and international standards bodies and the technical experts working on their committees before a standard is produced. National standards bodies still have an important role to play, contributing national views to European and international standardization efforts.

They also distribute drafts for comment and publish the final standard. In the UK, for instance, the British Standards Institution, BSI, publishes European standards with the prefix BS. Thus IEC 950 becomes EN 60950 becomes BS EN 60950. When BSI publishes a harmonized European standard, this may supersede an existing national standard.

If we look to our national and European standards organizations for concise and simple definitions then we will be disappointed: few (the cynic might say 'none'!) of these standards will present the information in a suitable way to the designer, and still fewer provide detailed examples of good design practice. To put these criticisms into context we must consider that what the standards writers are attempting is a very difficult (and arguably thankless) task. They must write clearly, use 'international' language that does not have any strange or conflicting usage and must write in only the most general terms.

If we compare the clarity of some of the official documents, such as the following abstracts, with which our countries abound then we must conclude that they have done a very good job.

> 'Technical Specification: A Specification contained in a document which lays down the characteristics required of a product such as levels of quality, performance, safety or dimensions, including the requirements applicable to the product as regards terminology, symbols, testing and test methods, packaging, marking and labelling, as defined in Article 38(a) of the Treaty of Rome and for products intended for human and animal consumption and for medicinal products as defined in Article 1 of Directive 65/65/EEC, as last amended by Directive 87/21/EEC.'
>
> 'Standard: A technical specification approved by a recognized standardization body for repeated or continuous application, with which compliance is not compulsory.'

On a lighter note, this next abstract comes from a 1970s EIA publication (the words will – I am sure – be forever timeless):

'A specification is a voluminous and painstakingly dry document designed to harass, hamper, and confuse the manufacturer, disturb the digestion and emotional stability of congressmen, gnaw at the very foundations of democracy, and provide simultaneous discrimination against both "big" and "little" business. It is written as a masterpiece of incoherence by a person who never saw the commodities specified . . ., and it costs more to write than the items described therein.'

The feelings contained in those words also convey how some people feel about European legislation, but as traders in the EU we really have no choice: if we are to survive – we must comply. Let us consider a few of the most common negative views and examine them in detail.

Why is there a need now?

'My company has been exporting to Europe, and much further afield. We've done that and built good markets without the need for this CE Mark nonsense or getting involved with technical standards. Why shouldn't I ignore this latest legislation, do nothing, carry on as before and save money.'

While we all may feel some sympathy for this view, and try hard to ignore these changes, there is no way that they will ignore us. There may be apparent, short-term financial benefits with the 'do nothing' option but there are huge commercial and personal risks.

Liability and compliance are important issues, heavily reliant on the use of standards. This is so important I want to spend a little time looking at how this situation came about, because if we understand what our politicians are trying to achieve, we can plan the best path for our business.

Why do we need standards?

The reason that some people ask the question 'Why now?' is because their experience of working with technical standards has been poor. They feel there was little need, because they were trading between fellow countrymen and had a perfect understanding of custom, needs and wants.

Now that there is a Single European Market, companies in all member states are subject to the same legal and technical requirements to supply safe products. However, there remains an underlying nationalism which could result in barriers to trade within the Single Market. If a product does not comply with the relevant standards, then member states have a perfect right to reject products coming from other member states. Just as compliance is a guarantee of free trade around Europe, non-compliance is practically a guarantee that you will be prevented from selling your product. Standards will soon be the technical currency that allows trade, not just in Europe, but around the world. I have described below some brief examples of countries whose businesses have benefited from the introduction of standards.

Japan

Many years ago, Japan introduced standards for, among other things, car rear lights. The specification required the plastic surface to be flat. However, all rear lights manufactured in the UK had 'Made in Great Britain' embossed on the outer surface. Before British cars could be sold in Japan they had to be modified by Japanese engineers to make them compliant. (For some reason the technical requirements were not available outside of Japan – or in English.) The embossed markings were filed flat, and the plastic surface polished (there were also many other technical modifications required). The cost of these modifications more than doubled the cost of imported UK cars and consequently decimated the exports to Japan.

In contrast, during 1992, I was visited by engineers from Fujitsu who were demonstrating equipment to prospective clients. The other part of their brief was to research what safety, EMC, and other standards their prospective customers expected, and needed to be in place. Within 12 months that equipment was commercially available and had all electrical approvals for the European, North American and Canadian markets.

North America and Canada

In many places it is illegal to connect any electrical appliance into the mains supply (even via a domestic plug) if it does not have a Canadian Standards Association (CSA) safety mark. Consequently, the majority of Canadian design engineers regularly design products to specific safety requirements, their work is regularly reviewed by CSA, and, naturally, these engineers usually get their design right. Hence the products are fit for the market, product rework and redesign costs are minimal, and time to market is short.

Similarly in many US states it is a criminal act to place any electrical appliance for sale if the appliance does not have an Underwriter's Laboratory (UL) safety mark. As with Canada the mandatory need to design and certify products has improved not only the safety of products, it has brought about improvements in product quality and the quality of design engineers.

Product liability

In 1987, the European Community Directive on Product Liability was enacted. This changes the onus in civil law to better protect the consumer. It removes the need for consumers to prove negligence in a legal action. If a manufacturer finds himself in court, his main defence is likely to be that he took all available measures to ensure that his product was safe. Note that this is a much greater burden than the requirements of criminal law. Should a manufacturer be taken to court against the Low Voltage Directive, he would only have to show that he took reasonable measures that a typical manufacturer might be expected to take. To defend a civil action, therefore, a manufacturer may have to go beyond the standards adopted against mandatory requirements and consider non-mandatory and emerging standards.

This legislation benefits the consumer who suffers damage to property or personal injury following the use of a defective product. The consumer has the right to take actions against all persons and companies in the European supply chain without any need to prove negligence. These actions may be taken against all parties simultaneously (joint and several) under civil law for claims for damages. There is a further right to initiate criminal actions against individuals such as designers or installers. This allows the individual to initiate action independently of the authorities. Considerations for deciding the result of criminal actions will be:

- Compliance of the product with relevant standards.
- The type and nature of safety information – is it adequate? Is it in an appropriate language?
- Compliance with minimum safety, EMC and other legal requirements.

EN 60950 and other standards

This book concentrates on EN 60950, the electrical safety standard relating to information technology equipment. This standard is certainly one of the most rigorous and any equipment that connects to IT apparatus, even if it cannot be regarded as IT equipment itself, it should meet EN 60950.

There are other safety standards relating to other product categories. You should always refer to the standard which is most appropriate to your product and ensure you have a copy of the most up-to-date version.

5 A detailed guide to EN 60950

Please have a copy of EN 60950 with you when you read this section of the book.

EN 60950 Standard reference	*Synopsis*
1.0	GENERAL
1.1	SCOPE Primarily intended for business equipment the 950 series of standards has been applied to a much wider range of equipments. (Personal note: Two possible explanations for this are: a general acceptance of the 950 series by an international audience; and the considerable attention to insulation and isolation which makes equipment certified to EN 60950 suitable for intensive use by non-technical operators and 'users'.)
1.2	DEFINITIONS
1.2.2.3	Continuous operation All equipment is assumed to be continuously operated and must be tested as continuous operation unless otherwise specified on the rating label.

When reading the applicability section I find it easier to start from the last section 1.1.3 and consider the equipment to which the standard does not apply. Following that look at section 1.1.2 which contains additional requirements for equipment such as equipment as above 2000 metres, where there are flammable atmospheres, medical equipment etc. Finally refer to 1.1.1.

Throughout this book – remember – if the beginning of the section is unclear, it may be more 'logical' to read the sections in some other order.

Most of the definitions are self-explanatory but there are a few which I would make some additional notes to.

For either short-term operation or intermittent operation. These definitions are covered in sections 1.2.2.4 and 5, both the allowable operation time and the period whereby the equipment must be rested must be specified on the rating label. (e.g. '25sec/2mins' means maximum operation time 25 seconds: equipment must not be operated until 2 minutes have elapsed since last operation.)

EN 60950 Standard reference	Synopsis
1.2.3	Equipment Mobility This section covers definitions for movable, hand held, stationary, fixed or equipment for buildings. These classifications are particularly important in how the equipment is going to be tested and how it can be used. It can also have a major impact upon the flammability ratings for the equipment. Some of the general considerations are the fire enclosure requirement for equipment weighing more than 18 kg, or equipment that is fixed must have the highest flammability rating (UL 94-5V). The fire enclosure requirement for equipment weighing less than 18 kg and that is portable or movable is UL 94V-1.
1.2.4	Classes of equipment – protection against electric shock
1.2.4.1	Class I equipment requires a protective earthing conductor.
1.2.4.2	Class II equipment does not require a protective earthing conductor wire.
1.2.4.3	Class III equipment receives its supply from a safety extra low voltage source and the equipment itself does not produce a hazardous voltage.
1.2.5	Connection to supply Pluggable type A equipment uses conventional, e.g. domestic style, plugs and sockets.
1.2.5.2	Pluggable equipment type B uses the industrial type, typically the IEC309 style connector.
1.2.5.3	Permanently protected equipment is designed to be installed and connected via screw terminals.

Guidance notes

If the equipment is nearly 18 kg or might need to be fixed to a building at some stage in its evolution, it is best to design for the higher flammable rating at the start than have to redesign later.

The positioning of rating labels will vary depending on its classification.

The stability tests requirements will also depend upon the equipment, its height and if it is to be fixed.

Make a note of all of the intended uses, installation and other options that are intended for the equipment during its product life cycle.

Read these notes and definitions carefully.

The subject is not difficult but a detailed understanding is necessary and is crucial to understanding the standard.

Minimum acceptable – one level of insulation plus safety earth ground – two levels of protection.

NB. Many parts of Class 1 equipment will use reinforced insulation (e.g. Plug sockets opto-isolators and pcbs). This situation does not affect the classification of Class 1 – it does, however affect type testing where reinforced insulation will be tested at higher voltages.

Minimum acceptable two levels of insulation basic – plus supplementary (or reinforcement) – two levels of protection. Reinforced insulation is equivalent to basic plus supplementary insulation and is considered to provide two levels of protections

These sources can still be a source of hazard as a result of high energy and high current. It is important to check all sections of the standard – especially energy hazards, flammability and mechanisms.

Maximum earth leakage current is 3.5 mA – there are other restrictions involving stored charge.

This allows earth leakage currents to exceed 3.5 mA and is the only universal polarized type of connector.

There are requirements to warn the service or installation engineer, a few of the items include: disconnect devices; short-circuit protection; earth leakage; mechanical features and many more.

EN 60950 Standard reference	*Synopsis*
1.2.6	Enclosures There are three special definitions of enclosures. They are covered by sections 1.2.6.2 to 1.2.6.4 and cover: fire enclosure; mechanical enclosure; electrical enclosure.
1.2.8	Circuit characteristics
1.2.8.1	Primary circuit This can generally be considered as anything which is connected to the mains supply or that has mains potential connected to it.
1.2.8.2	Secondary circuit This applies to any circuitry which is not directly connected to primary power. Usually it relates to the output of transformers or some other isolation device.
1.2.8.5	Safety Extra Low Voltage (SELV) circuit.
1.2.8.6	Limited Current Circuit In terms of safety this is similar to SELV because it is not possible to draw hazardous current under normal or single fault conditions.

Do make sure that publicity material and instructions use the phrases 'fire enclosure', 'mechanical enclosure' and 'electrical enclosure' in their appropriate context. The same comment applies to all definitions within the standard, in particular, the words 'operator' and 'service personnel' must never be confused in documentation and literature. (An operator must be protected from any single fault and must be given two levels of protection – a warning notice does not generally count as a 'level of protection').

The phrase fire enclosure is a special term applied to parts of the equipment which are designed to minimize the spread of fire and comply with later sections of this standard.

Mechanical enclosure is the special phrase used to describe parts of the equipment intended to prevent injury to an operator or service engineers due to mechanical or other physical hazards.

Electrical enclosure is the special term to describe any part of the equipment which prevents contact with parts at hazardous voltage or energy levels.

It is vital that they are well understood by all associated with the project.

It is a good idea to highlight all primary circuits in red and all SELVEL circuits with a green highlight marker. This will show areas where creepage and clearance are vital to safety and help identify areas of basic or reinforced insulation. It will also show some critical components that may have been overloaded.

The use of isolation devices for a secondary circuit does not guarantee safety.

Secondary circuits may contain hazardous voltages/currents or energies and have only basic insulation. One level of protection from a hazardous source is not enough.

The output from an SELV power supply will provide protection from hazardous voltages under normal and single fault conditions. Typically this is a voltage of less than 60 V d.c. – there is generally no limit to the current that can be drawn from SELV supplies, hence they can be a potential energy hazard.

The only circuits that an operator may touch are Safety Extra Low Voltage Energy Limited circuits (SELVEL) or limited current circuits.

EN 60950 Standard reference	Synopsis
1.2.9	Insulation Generally, the equipment should be safe and not present any hazard under single and likely fault conditions. This section contains detailed definitions.
1.4	GENERAL CONDITIONS OF TEST Section 1.4.1 exempts aspects of the design from safety testing if safety isn't affected.
1.4.2	This defines the test within the standard as type tests which means that unless otherwise stated, they do not need to be performed during routine production on 100% of products.
1.4.3	Defines how tests should be carried out at subsystem level.
1.4.4	Requires that tests be carried out in the most unfavour- able combination of parameters but within the manu- facturer's operating spec. That means supply voltage frequency adjustment of thermostats which are operator accessible and which can be set without the use of the tool, unless there is a tool provided with the equipment or by the supplier of the equipment for use by the operator.
1.4.5	Deals with specifying and establishing the most unfavourable supply voltage.

Designers should read section 5.4 dealing with abnormal testing that their designs will be subjected to before rather than after they start designing. In particular the passage from section 5.4.6.

In addition to these definitions it may be helpful to picture a mains cable where the hazardous voltages are being carried on the copper conductors protected by blue and brown basic insulation (which the operator may not touch). These basic insulators are covered by supplementary insulation and together form a double insulation system. If you now look at the body of a plug which has only one piece of insulating material between the live conductor and the operator, that plastic insulator provides reinforced insulation.

A sign reading 'equipment is tested by applying any condition that may be expected in normal use and foreseeable misuse' would be well placed on the door to any laboratory or design house, as a constant reminder.

When completing the design safety checklist and the safety plan remember to identify non-applicable sections and to make an explanatory note as to why they are not applicable. These explanations will be vital to the company if it is called upon to substantiate its compliance claims.

Type tests are usually performed on one unit only. They should be considered to be destructive and usually include: Mechanical; Earth Bond; Hi Po; electrical; thermal and abnormal tests.

Type tests do not need to be performed during routine production on 100% of products. Production testing usually includes Hi Pot and Earth Bond.

1. Tests may be conducted on components and materials before committing these items to the parts list.
2. Component or sub-assembly bench tests.
3. Unpowered tests.
4. Powered test.
5. Abnormal and destructive tests.

Do not make assumptions; there are many occasions when reducing the supply voltage can identify a non-compliance (e.g. fuses do not always blow and components catch fire under single fault conditions, the current inputs to switch Mode Power Supplies will increase at low voltages and input filters may overheat). We are only assured of the worst condition by practical measurements.

EN 60950 Standard reference	Synopsis
1.4.6	Deals with identifying the most unfavourable frequency supply.
1.4.7	Details how changes in test temperature, and ambience temperature should be dealt with during the series of tests.
1.4.8	Deals with measuring the temperature of copper windings using either a thermocouple or 'the rise of resistance' method.
1.4.9	Deals with examining the input current. If the equipment is to have other units plugged to it or provide outlet sockets in operator accessible areas then the maximum shall be drawn from those sockets to the limit of the type of connector socket plug that is provided.
1.4.10	Deals with conducting liquids.
1.4.11	Deals with the bandwidth of measuring equipment.
1.4.12	Deals with applying simulated faults or abnormal operating conditions. In particular it covers consequential faults. These can result following the introductions of faults or abnormal operating conditions. For instance, if a particular fault creates overheating in wiring looms, and that damages insulation in other parts of the circuit. The secondary and all consequential faults are considered as part of the initial, single fault. If the consequential fault results in an equipment failure then the equipment fails the test. Before the next (single) fault is introduced in the sequence of simulated faults, or abnormal testing, the equipment must be made fully functional and to its normal production standard (e.g. repaired). The text of the standard suggests examining circuit diagrams and component specifications to determine which fault might reasonably be expected to occur. In

Guidance notes

Test for all frequencies. Remember that iron circuits (e.g. transformers) optimized at 60 Hz will saturate and overheat if run at 50 Hz. Test at both frequencies. If specified as 50–440 Hz then test must be over the full frequency range.

The *rise of resistance method* is the simplest method of measuring copper temperature in windings, transformers, and motors.

A 2-pole change-over switch will allow instantaneous switching between test and sense.

Unless connectors are of a specialized type and the additional equipment connected to them is always provided with the equipment under test, it is essential to place a label near the connector specifying the maximum load current that can be drawn by it, e.g. a British 13 A socket would be connected to an external load drawing 13 A for the duration of the tests – unless there was a label specifying that a lower maximum current may be drawn (e.g. '230 V 5 A max').

This section specifies the tests and design criteria to deal with overfilling of liquids. Remember that if there is a hazardous spill it is our responsibility to tell the operator what to do.

Avoid using oversensitive measurement equipment.

Check flying leads in all secondary circuits which might break or become detached. These must not be able to contact parts in primary or hazardous circuitry.

Check all conductors carrying hazardous voltages or currents which might break or become detached. These must not be able to contact parts in SELV. If *any* piece of basic insulation fails there must be two levels of protection between hazardous voltages and SELV.

When using crimp connectors specify the type which crimp on to the copper cable and also on to the outer insulation of the wire. This type is not likely to release the inner conductor under single fault conditions; also ensure that these type of connections are of the locking type so that when they are mated they lock to the receptacle and cannot be removed inadvertently, accidentally or by vibration. This selection of connectors will reduce the likelihood of faults to be investigated. Alternatively solder the connections and use heat shrink sleeving and tie wraps to hold them in place. Wires which can become detached and create a possible hazard should be retained with tie wraps, the positioning and location of these tie wraps will form part of the safety critical definition of the equipment.

EN 60950 Standard reference	Synopsis
	particular it suggests short-circuits and open-circuits of semiconductors and capacitors; faults causing continuous dissipation in resistors designed for intermittent dissipation; internal faults in integrated circuits causing excessive dissipation; failure of basic insulation between current-carrying parts of the primary circuit and the following: accessible conductive parts, earthed conductive screens, parts of SELV circuits, parts of limited current circuits.
1.5	COMPONENTS
1.5.1	This requires all safety critical components to comply with the relevant IEC component standard or the requirements of EN 60950. It also requires components that connect between SELV and either ELV or parts of hazardous voltage to comply with requirements of section 2.3. This is to ensure that the SELV integrity of the system is not violated.
1.5.2	Evaluation and testing of components The most significant impact of this clause is that components not certified against the relevant IEC standards (e.g. without a safety mark) must be tested fully for compliance to the IEC standard. This is potentially a very expensive path and implies continuous testing of each batch during the manufacturing life of the product. Thermal controls are tested in accordance with section K of the standard.
1.5.3	Requires transformers, and isolating transformers to be tested in accordance with the requirement of annex C of the standard.
1.5.4	Requires high voltage components operating at voltages exceeding 4 kV to have flammable class of V-2 or better or HF-2 or better or to comply with section 14.4 of IEC 65:1985.
1.5.5	Deals with interconnecting cables supplied with the equipment and including them within the full safety requirements of this standard.
1.5.6	This details the selection of X-capacitors suitable for connection between phase conductors or phase and neutral of the primary mains supply.

When moving wires to their extremes check also the proximity to hot components such as wirewound resistors, inductors, chokes and other wound components.

This subject is a major topic and will, if demand is sufficient, be the subject of further publications.

Safety critical components can be summarized as those that perform a safety function or whose lack of integrity could create a potential hazard. For instance, for all components and parts in primary circuitry, primary hazardous voltage should be listed as safety critical components; any interlock or other similar device would also be defined as a safety critical component: labelling or warnings are also safety critical.

Very substantial cost savings can be achieved by identifying safety critical components and ensuring that these are correctly specified to their IEC or other relevant safety standard.

This is particularly relevant when making the decision to design or purchase a power supply for a piece of equipment: the choice to use an approved power supply will reduce the need for much investigation and testing of the system.

The design of safety isolating transformers is a specialist task and should not be taken lightly.

When we supply external cables – looms or power cords – we must ensure that they satisfy the flammability, insulation and dimension requirement of the standard.

Capacitors connected across primary circuits should be 'X1' and mounted to show compliance with IEC 384-14:1981.

Other options are allowed but ensure that the components are marked

EN 60950 Standard reference	Synopsis
1.6	POWER INTERFACE
1.6.1	Allows the steady-state input current from equipment to exceed the rated current on the rating label by up to but not exceeding 10 per cent under normal load conditions. It also defines the test methods and test conditions under which input current is to be measured.
1.6.2	Allows hand-held equipment to have a rating voltage of up to but not exceeding 250 V.
1.6.3	Requires the neutral conductor to be insulated from earth and other parts of the equipment as though it were a phase conductor. In particular any component connected between the neutral conductor and earth must be rated as though it were a phase to neutral voltage.
1.6.4	Deals with equipment intended to be connected to IT power systems.
1.6.5	Defines the minimum supply range that equipment must be designed to operate over.
1.7	MARKING AND INSTRUCTIONS
1.7.1	Power rating This section deals with the mandatory rating information that must be displayed on all mains connected equipment. The rating label must include the rated voltage or voltages or the rated voltage range or ranges, the frequency of the supply, the rated current, the manufacturer's name or trademark or identification mark, the manufacturer's model or type reference, the symbol for Class II construction if it is a piece of Class II equipment. This section requires the manufacturer to make all

Guidance notes

– accompanied with some form of 'certificate of conformity'. I am told that there are counterfeit components on the market – therefore be careful and observe 'due diligence'!

Normal load conditions would be defined on the rating label in terms of intermittent operation as well as the user documentation and perhaps also the installation documents. It is quite acceptable to give equipment that draws 100 mA a rating of 1 A – we should not rate equipment excessively high but it makes good commercial sense to round up to values that allow pre-printed rating labels to be made.

Hand-held equipment may generate voltages greater than 250 V.

This is to accommodate:

1. Non-polarized European plugs where live and neutral can be transposed during normal operation of the equipment.
2. Short-circuits between phase and neutral that could raise the voltage on the neutral line to over 100 V.

In this context IT does not mean Information Technology but refers to power distribution systems that have no direct connections to SEG and where accessible conductive parts are earthed.

It hardly needs stating that it is good design practice to design equipment that will exceed the tolerance of +10% −10% for 230 V or 400 V, or +6% −10% for any other voltage.

Many cases of non-compliance result from incorrect labelling – and many more could be addressed by appropriate labelling.

Although not difficult this section is usually one which most often causes non-compliance.

At first we will consider where the label should be mounted. If the equipment may be installed by the operator the label is best placed near the mains power inlet where it will be visible after the equipment is installed. If the label cannot be placed on the outside of the machine but must be placed in a service area or behind a door or cover then there must be a label showing the location of the rating label ('Rating information behind this panel').

The following gives an indication of the notations that can be used. The examples are not exhaustive but are intended to indicate good design practice and limitations.

EN 60950 Standard reference	Synopsis
	necessary safety instructions and safety information available to avoid the introduction of hazards during operation, installation, maintenance, transport or storage.
1.7.2	Safety instructions
1.7.3	Short duty cycles This relates to equipment intended for short-term or intermittent operation. The basic requirement is that equipment intended to be used for a short period of time or for an intermittent duty cycle must either be constructed so that its operating time is limited by its construction, or the rated operating time must be clearly specified on the equipment.
1.7.4	This section deals with mains voltage adjustment for equipment intended to be connected to multiple voltages or frequencies and which is intended to be set by the service or installation engineers.

Guidance notes

- '230V' nominal European supply range.
- '220/240'. Note that the marking means either 220 or 240 and the mains voltage is operator selectable. (This is usually an error on behalf of the manufacturer – it is unusual to find equipment that is incapable of operating over the European range.)
- '110/230' operates on either the 110 supply range or the European 230 range. Voltage selection can be set be the operator, usually with a switch.
- '110–230' operates over the two ranges and anywhere in between. (e.g. from about 100 V to 254 V).

Consider local language or international symbols.

A maintenance warning could be fuse replacement information, disposal of batteries etc.

Installation warnings should include warning of high earth leakage or where equipment is likely to be interconnected and the resultant earth leakage current may be high (e.g. if five computers are plugged into a single distribution strip the potential earth leakage current may be more than three times the legal maximum!)

Installation may include fitting additional circuit breakers or disconnect devices into the equipment wiring.

Operating warning would cover all aspects of the equipment that could become hazardous during expected normal use or foreseeable misuse.

The language for these instructions and warnings is dealt with in section 1.7.14.

These include written instruments and notes or warning labels on the equipment or its packaging.

Typical solutions are to provide timers so that the equipment can only be operated for specific time internals, thermal switches on the components which become hot or over-dissipate or potentially over-dissipate and rating information on both the rating label and near the operating switches.

Design considerations must include the failure of the timing devices and thermal switches to ensure that a single failure does not create a hazard within the equipment or to its operators.

An acceptable marking for equipment that can be operated for 20 seconds every 5 minutes is as follows:

'20 sec/5 min'

This section applies only if the mains voltage adjustment is intended to be carried out by service engineers or installation engineers.

It does not apply to equipment which is supplied for a specific voltage (even if it contains a power supply which has been factory set to that voltage).

EN 60950 Standard reference	Synopsis
1.7.5	Power outlets This requires any standard power outlet (e.g. domestic mains socket or a socket conforming to IEC 83) which are accessible to the operator to be clearly identified with the maximum load that can be drawn from that outlet.
1.7.6	Fuses Where fuse holders are operator accessible there must be a clear marking giving the fuse rated voltage, current, *and* special characteristics such as time or breaking capacity. If fuses are not operator accessible and are in service access areas they may be marked F1, F2 etc. and the fuse current time and voltage information can be given in a cross-reference in servicing instructions.
1.7.7	Wiring terminals
1.7.7.1	Deals with wiring terminals. The particular requirement is to identify the protective earthing conductor (the safety earth ground) with the IEC 417 No. 5019 (symbol of an earth symbol in a circle). The purpose of this is to clearly identify the primary earth point on equipment.
1.7.7.2	On three-phase equipment where incorrect phase connection could cause overheating or other hazards,

The instructions recommended within EN 60950 section 1.7.4 may be contained on a temporary label which can be removed after installation.

If the operator can adjust the mains voltage then the equipment must be tested with the settings for the wrong voltage and frequency as part of its abnormal series of tests.

The alternative to this would be to test the product at the maximum rating for the socket outlet.

It is unlikely that the designer will know what fuses to fit into the equipment until abnormal and fault testing have been completed (e.g. 250 V 2AT, 250 V 1AF – 'T' = time delay, 'F' = fast blow, 'H' = high breaking capacity).

There is no need to identify standard (glass) fuses which can usually only break fault currents of about 35 A.

Time characteristics are vital; incorrect, or inadequate marking will increase the risk of fire.

NOTE FOR US DESIGNERS:
UL recognized fuses will blow at their rated current: European fuses will pass their rated current indefinitely. Therefore you will need to retest (and perhaps respecify).

Before deciding upon the mains inlet fuse rating it is very useful to perform an unscheduled unnecessary switch on test. Turn the equipment on and off at one-second intervals to ensure that the mains inlet fuse does not blow under 'normal' circumstances.

Above all ensure that all fuses fitted to the equipment carry an appropriate safety approval mark.

The conventional British 13A plug should normally be fitted with either a 13 A fuse or a 3 A fuse.

This section of the standard deals with potential non-compliance when using terminals or connectors which are pre-marked with the incorrect symbols. It should be noted that these requirements apply even to equipment which is pre-wired with a non-detachable power cord.

Make sure that all markings cannot be removed or interchanged and are not obscured when wiring is installed.

Make sure that earthing points are obvious and not slackened, in error, by service personnel trying to remove access panels.

EN 60950 Standard reference	Synopsis
	individual phase connections must be clearly identified and referenced in the installation instructions to ensure that they are connected correctly.
1.7.8	Controls and indicators
1.7.8.1	States the general requirement that where safety is affected all indicators, switches and controls should be identified or placed so that their function is obvious and that indications used to identify these devices should wherever possible use symbols that do not require language or translation.
1.7.8.2	Take note of IEC 73 and the section on colour.
1.7.8.3	Requires the on/off and standby symbols of a vertical line for 'on', a circle for 'off' and a line within a circle for push-button type switches. It also covers the standby marking which is a broken circle with a line through the break. These symbols are defined in IEC standard 417 Nos 5000, 7, 8, 9 and 10.
1.7.8.4	This refers to controls which have more than one active setting.
1.7.8.5	This is a simple requirement that markings and indications for switches and controls should be located near the switch or control, or that the markings and indications should be obvious as to which control they apply to.
1.7.9	Isolation of multiple sources This applies to equipment where more than one mains inlet or where a mains inlet and an additional supply of hazardous energy or voltage is present. The requirement is to place an obvious warning for service personnel to advise them to disconnect hazardous supplies.
1.7.10	Deals with IT power supplies (IT power supplies have no direct connection to earth). The requirement is for equipment instructions to state clearly the equipment's suitability for use with IT power systems.

Guidance notes

If in doubt fit a Safety Earth Ground label to both sides of panels to indicate the function of grounding hardware.

The important feature here is where safety is affected.
If the control or switch has no safety effect then it is not included within this section of the requirements.

Within Europe we may now use red lights provided it is obvious that they are not warning. Personally I still prefer to use red as a warning only.

Wherever possible make the on/off positions of switches obvious. Avoid using neon lit switches which do not have the on/off symbols either on the switch or on the equipment. The reason being that if mains supply to the equipment fails or is turned off there is no indication of the on/off switch position. Similarly try to avoid the use of the words on/off and standby; try to use symbols.

The control may relate to an input or an output or could refer to a variety of operating conditions. This section requires the inactive off state of the control to be marked with a 0 and other higher numbers (e.g. 1, 2, 3, 4, 5, etc.) to be used to indicate upper states.

The use of a diagram attached to the equipment is useful. However, ensure that languages are appropriate for the country of use.

Acceptable warning is black lettering on a yellow or amber label reading 'Caution disconnect 2 (3 or more) power cords before servicing (insert the correct number of power cords for the equipment etc.)'.

If the equipment has been modified after use on an IT power system, it should be clearly marked : – 'This equipment has been modified for use on a IT power system only.'
Acceptable notes are 'This equipment is not suitable for connection to

EN 60950 Standard reference	Synopsis
1.7.11	Protection in building installations This section relates to pluggable equipment type B (connected by an industrial IEC 309 or similar connector) or equivalent that is permanently connected and which requires protective devices to be installed in the building. The protection in the building may be either for overcurrent protection or for short-circuit protection or to provide a disconnect facility.
1.7.12	High leakage current This requires any equipment with an earth leakage current exceeding 3.5 mA to carry a warning label. For full details of earth leakage refer to section 5.2 where the whole matter is dealt with. The significance of earth leakage current is that under certain single fault conditions (e.g. the earth connection within a domestic plug becomes detached) the entire earth leakage current may pass through the operator's body. Table 17 of EN 60950 defines the maximum limits of earth leakage for various classes of equipment and their methods of connection.
1.7.13	Thermostats and other regulating devices must have an indication of the direction for their minimum and maximum adjustments.

Guidance notes

an IT power system'. Alternatively, detailed instructions explaining any modification that would be necessary at installation must be included in the instruction manual.

This is not as simple as it appears, there are many implications to this section:
All details must be included in servicing instructions.

The characteristics of the device must be specified in the installation instructions, this must include voltage current and time characteristics. The breaking characteristics must also be specified in the installation instructions.

If using circuit breakers check the maximum breaking current: any fault current in excess of this will not be interrupted by the circuit breaker and may result in a fire with the equipment or overloading of the building wiring.

Therefore, make a clear statement in wiring insulation instructions that the equipment is internally protected for fault currents of up to (insert here the maximum specified fault current) fault currents. Fault currents in excess of this must be 'protected by the building supply in accordance with local and national wiring regulations'.

Ensure that fault testing is carried out using building protection systems as specified in the installation instructions. This is to verify that equipment wiring is adequate for fault currents and that insulation will not suffer (adiabatic) heating and sustain damage.

It is recommended that a typical earth leakage figure should be included in the equipment instructions – this will indicate any limitations in the number of equipments that can be connected using a mains distribution strip.

PERSONAL APPEAL:
I have had a 'thing' about earth leakage ever since I measured 19.7 mA on equipment fitted with a conventional mains plug.

The equipment comprised a number of individually powered (highly suppressed) units mounted in a rack. Nowhere in any of the individual data sheets was there any information to give the designer guidance in determining the final leakage current.

Please do not include a statement like 'Less than 3.5 mA' that really does no one any good.

Remember, if you include helpful information with your product your customer will like you and is more likely to buy more.

Alternatively, your customer could be someone like me!

A simple plus sign or minus sign is acceptable. The symbols from IEC 417

EN 60950 Standard reference	Synopsis
1.7.14	Language All safety related instructions and equipment markings intended for the operator must be in an acceptable language for the country in which the equipment is to be installed and used. Documentation and instructions or warnings intended purely for service personnel may be in English only. However, these warnings should follow the same format of the nature of the hazard and how to avoid it. (E.g. Electrical hazard; this unit is powered with an uninterruptible power supply (UPS). Disconnect UPS before servicing).
1.7.15	Durability Obviously markings must not rub off or fade with age. This section deals with test methods to ensure their durability.
1.7.16	Removal of parts It must not be possible to remove parts of the equipment and to replace them such that markings required by the Standard can become misleading.
1.7.17	Lithium batteries The reason for this section is that lithium batteries can explode under certain circumstances.
1.7.18	Operator access with a tool – this section applies only if the operator requires a tool to gain access to an operator access area (e.g. use of a screwdriver to release an operator

It is for this reason that using international symbols is the preferred way of indicating hazards or control functions.

When writing safety instructions it is also best to follow the format: the nature of the hazard; and what operations must be taken to avoid the hazard. (E.g. Hot surface, do not touch, electrical hazard, no user serviceable parts within this enclosure.)

Note that service warnings can generally be in English.

The only exception (for the 950 series of standards) is Canada where they *must* be in English and French.

The test requires rubbing the marking for 15 seconds with a cloth soaked in water and for a further 15 seconds with a piece of cloth soaked in petroleum spirit. The standard defines the content of the petroleum spirit for this test.

I carry a tin of petroleum lighter fuel for informal testing and found that it has achieved similar results to the materials specified in the standard.

It must not be possible to remove parts of the equipment and to replace them such that markings required by the standard can become misleading.

Do not put the product model number, serial number, warning or other information on removable parts.

When placing switches, controls and their labels on removable parts check that the removable parts cannot be reassembled either to hide the labels or to change the orientation between the labels and the controls or switches to which they relate.

If lithium batteries are placed in operator access areas where they will be replaced by operators, then there must be a warning notice including the text within the standard. (E.g. Caution danger of explosion if battery is incorrectly replaced. Replace only with same or equivalent type recommended by the manufacturer. Dispose of used batteries according to the manufacturer's instructions.)

Obviously this warning must be in the appropriate language for the country of installation and use, and user instructions must include safe disposal instructions. If the lithium batteries are in service access areas and will be replaced by service personnel, then the warning may be in English in the servicing instructions. Once again, safe instructions for disposal must be included.

Take particular caution when using quick fasteners for service use, these can easily be removable with pressure.

If the operator is required to use a tool specify that tool exactly in the

EN 60950 Standard reference	*Synopsis*
	access panel). The standard requires that either the operator's tools cannot be used to gain access to other parts of the equipment or that any other part of the equipment to which he could gain access with the tool is marked with an electrical shock hazard. This is the lightning flash within a triangle; the reference for that mark is ISO 3864 No. 5036.
2.0	FUNDAMENTAL DESIGN REQUIREMENTS
2.1	PROTECTION AGAINST ELECTRIC SHOCK AND ENERGY HAZARDS
2.1.1	This section deals with the general requirements for protecting operators and service engineers from energy hazards and against contact with TNV (telephone network voltage) circuits.
2.1.2	Identifies protection against energy hazards and protection against contacts with telephone network voltage (TNV) circuits. The operator may touch SELV bare parts in limited current circuits and extra low voltage (ELV) circuits provided that the insulation meets the criteria specified in section 2.1.3. The operator must not have access to bare parts at ELV or circuits at hazardous voltage or circuits containing energy hazards. The operator must not have access to operational insulation or basic insulation unless special criteria are met, refer to section 2.1.3. The operator must not have access to unearthed conductive parts which are separated from ELV or hazardous voltages by operational insulation or basic insulation only.

operator instructions and test all other potentially hazardous areas for access using that tool.

If in doubt apply the lightning flash ISO 3864 No. 5036 to surfaces containing hazardous voltages or energies.

The considerations within this section of the standard cover the protection of an operator for likely single fault conditions. (E.g. Failure of basic insulation, and the protection of service engineers from unexpected hazards, e.g. unintentional contact with hazardous voltages.) If the reader is not already doing so they should now be cross-referring section by section to the main body of the standard.

In the context of operator contact there is no compromise to the principles of SELV construction and limiting operator accessible currents and energy to acceptable limits.

If there are external electrical connections to the equipment the cost of compliance testing can be greatly reduced by incorporating an approved SELV power supply.

If using limited current circuits ensure that all normal and single fault conditions are considered (e.g. short-circuit of any one resistor, components becoming detached, creepage and clearance distances etc.)

When wiring ELV circuitry consider that a single failure of basic insulation may cause the ELV circuit to be at mains potential: hence it is advisable to use wire rated for mains potential when wiring ELV circuits (this will considerably reduce the amount of assessment and testing that must be conducted, it may also reduce the amount of earth connections to operator contactable panels).

ELV circuits may under single fault condition become hazardous; operational insulation does not provide adequate safety for operators or service engineers; unearthed conductive parts could become live following a failure of operational or basic insulation unless they are adequately earthed.

The use of good traditional engineering practice of containing wires in looms will usually avoid stray wires contacting operator accessible unearthed conductive parts. When carrying out the tests identified within this section, the objective should be to attempt to contact hazardous or other parts and to note the design features that prevent hazards from occurring. These mechanisms are safety critical aspects of the design and

EN 60950 Standard reference	Synopsis
2.1.3	Permits operator access to internal wiring at ELV.
2.1.3.1	The wiring must not be subject to damage or stress, not need to be handled by the operator and fixed so that it does not touch unearthed accessible conductive parts. Also it must have an insulation of greater than the levels specified in section 2.1.3.1.
2.1.3.2	Restates the requirement that if wiring carrying hazardous voltages must meet the requirement of reinforced insulation of the combination of basic insulation plus supplementary insulation.
2.1.4	This allows parts at hazardous voltage to be exposed to service engineers provided that they are located and guarded such that unintentional contact is unlikely during servicing operations. This will obviously depend upon the way in which service personnel will need to gain access to or past these potentially hazardous areas.
2.1.5	Requires that there is no energy hazard in the operator access area. The checks for compliance here involve test fingers as described in the text of the standard.
2.1.6	The clearances behind earthed and unearthed conductive parts of an enclosure must not be reduced by applying a force of up to 250 N if that would result in an energy hazard.

should be recorded and controlled during manufacture. Typically these will include the location of cable restraints, the fitting of additional electrical sleeving, the routing and fixing of looms. Take particular care when using self-adhesive cable restraints: these may fail under high temperatures and allow cables to move causing non-compliance.

My personal preference would be to avoid this option by fitting additional (supplementary insulation) or routing with wire internally.

Having made that point, if you must use this option – make sure that the wire has the correct thickness of insulation for its maximum voltage; that it is not movable – is not subject to any stresses – and that the operator will make only occasional contact with it.

Note to US designers and importers of US products: UL1950 permits basic insulation to contact unearthed conductive parts as a 'D3' deviation.

This deviation does not apply in Europe hence any product using the deviation will be non-compliant (and potentially subject to recall if sold) within the EU.

There is no relaxation for potential energy hazards in service areas. All bare parts that involve a potential energy hazard (20 J or greater) must be located enclosed or guarded to eliminate the possibility of unintentional bridging by conductive tools or materials during servicing.

Take particular attention of ELV and hazardous voltages which are not obvious (e.g. unearthed heat sinks in switched mode power supplies could be marked with a warning label 'unearthed live part').

The simplest way of avoiding the need to consider energy hazards for servicing is to ensure that all capacitors are discharged before covers are removed. This can be covered with a simple warning in the service manual and on the equipment. First measure how long it takes for capacitors to discharge until the energy stored is less than 20 J. The warning notice could read as follows. 'Remove power and wait x minutes before servicing' (where x is the time for capacitors to discharge). A note on the equipment could read, 'See servicing instructions before removing cover'.

The limits here are quite complex and depend on the voltage available. These levels are clearly defined within the standard.

Leave a good clearance between metal parts of the enclosure and parts at ELV hazardous voltage or containing an energy hazard. If space does not permit this consider including a suitable insulating material, a piece of fibreglass board could be appropriate for this.

EN 60950 Standard reference	Synopsis
2.1.7	Shafts of all operating knobs, handles, levers and any other control shall not be conducted to hazardous or ELV circuitry.
2.1.8	This relates to conductive handles, levers, control knobs and the like which can be manually moved in normal use and which are earthed only through a pivot or a bearing. The requirement is that they should be treated as an unearthed metal part and separated from hazardous voltages by the equivalent creepage – clearance distance of double or reinforced insulation.
2.1.9	Prohibits connecting the conductive cases of capacitors operating in ELV or hazardous voltages to unearthed conductive parts in operator access areas. It requires that cases should be separated from these parts by supplementary insulation or by earthed metal.
2.1.10	Requires that equipment when disconnected or unplugged must not present a risk of electric shock from stored charge. EMC filters of up to 0.1 μF, or any filter or capacitor marked greater than 0.1 μF, must have some lead resistor which will provide a discharge time constant of less than 1 second for pluggable type A (domestic connectors) or 10 seconds for permanently connected equipment or pluggable equipment type B (industrial, IEC 309 connectors etc.).

Guidance notes

If plastic is used check that its flammability is appropriate.

This applies both to operator access areas and service access areas. In operator access areas the reasons for this are obvious. In the service access areas such a situation could create an unexpected hazard for the service engineer.

NB. The use of insulation screws or similar small items to prevent contract will be considered non-complication if replacing them with conductive screws would produce a hazardous condition.

Ensure that all looms carrying hazardous voltage are either held well away from conductive handles and parts, that the wiring is rated at the equivalent of reinforced insulation, or is covered with supplementary insulation. This solution is generally more easily implemented than other methods of reliably covering operator accessible metalwork with reinforced insulation.

Take care with retaining screws for capacitor fixing clamps. There are several opportunities for non-compliance. Consider a single fault which allows the capacitor mounting bracket to be raised to ELV or hazardous voltage. The screw holding this bracket may also be raised to hazardous voltage or ELV: if the head of the screw is in an operator accessible area then the equipment will fail to comply (this is because screws may be used for one purpose only for earthing or for fixing – they may not have dual functions).

The use of non-conductive screws may also present an opportunity for non-compliance since if these are incorrectly replaced with conductive screws the equipment would become non-compliant.

The use of two screws holding the capacitor clamp in place would be compliant since one screw would be declared and identified as the earthing connection and the other screw would be designated the fixing device.

Equipment in which this requirement can cause design problems are switch mode devices and equipment containing motors which can regenerate hazardous voltages before they have stopped rotating.

Take particular care where large capacitors are charged directly from the mains or hazardous supply. Can controls, protection devices or interconnections allow these capacitors to be charged prior to disconnecting from the power source? If this situation arises it is possible that (with the mains on/off power switch 'on') that the exposed pins of the plug at the end of the power mains cord will allow operator or service access to hazardous and high energy sources which have only a bridge rectifier as protection – this clearly fails to provide adequate creepage, clearance, or distance through insulation to provide the necessary operator or service protection.

EN 60950 Standard reference	*Synopsis*
2.2	INSULATION
2.2.1	Introduces the basic principles of insulation as solid or laminated insulating materials having adequate thickness (distance through insulation) and creepage distances over their surfaces. Alternatively, adequate clearance through air is acceptable to provide insulation.
2.2.2	Excludes the use of hydroscopic (materials which absorb water or moisture) for use as insulation and requires the designer to take account of electrical, thermal, mechanical and frequency of the operating voltage and the working environment.
2.2.3	This defines humidity treatment or insulating material and enclosed or sealed parts referenced in section 2.9.6.
2.2.4	Requires equipment to comply with electric strength requirements of section 5.3, with the creepage clearance and distance through insulation requirements of section 2.9 and with the heating requirements of section 5.1.
2.2.5	Requires the designer to consider the application and also working voltages when determining test voltages to be applied to insulation.
2.2.6	Describes applications for operational basic supplementary reinforced or double insulation. Between a primary power conductor and an earthed

Guidance notes

This standard allows only these two methods of insulation: it does not acknowledge the use of semiconductor devices or liquids to provide adequate protection to operators or service engineers.

Plastic insulating materials may not be suitable in applications where the insulation may be pinched or subjected to high contact pressures and cut through. Materials like PTFE are subject to cold flow and looms that are tightly bound may, over many years, become short-circuit.

When selecting terminal strips or similar electrical components choose products with appropriate safety approvals – they will have been subjected to thermal and other tests which should cover most equipment environments, hence reducing the need for detailed component testing.

Note also sections 4.4.4 for flammable 'parts within 13 mm of electrical arcing parts', 'within 13 mm of high temperatures', and section 1.5.4 for components operating 'in excess of 4 kV'.

Unapproved insulation is tested subject to clause 5.3.2 or annex C.3 and should (where required) be preconditioned by the humidity treatment in section 2.2.3.

The use of approved components which have been proven to meet requirements will reduce the need for these tests.

Section 2.3 electric strength requirements will be a series of tests performed on the equipment. Section 2.9 creepage clearance and distance through insulation are a series of design criteria which must be followed. Section 5.1 heating requirements defines the maximum temperature rise (based on an assumed maximum ambient temperature of 25 °C: where operating temperatures exceed 25 °C then these maximum temperature rises must be reduced). These temperature rise limits relate to electrical insulation, power supply cords, terminals, components and operator contactable handles, knobs, grips, external services of equipment and parts inside the equipment. These are fundamental design criteria which form part of the design process.

These are described at length in the following two sections.

Operational insulation provides no level of safety whatever.

Basic insulation provides a single level of protection. Supplementary insulation performs a similar function to basic insulation in that it provides

EN 60950 Standard reference	*Synopsis*

screen or a core of a primary power transformer. (Please note that this is a minimum requirement and many transformers are produced with reinforced insulation between these two circuit elements.)

As an element of double insulation. (As we have stated before, double insulation comprises two parts, basic insulation and supplementary insulation. The simple example is a mains conductor where the hazardous copper conductor is surrounded by a brown basic insulator (or black in North America) which in itself is surrounded by an outer sheath to the cable).

Supplementary insulation
In general supplementary insulation will be used between an accessible conductive part and a part that could be raised to a hazardous voltage in the event of failure in basic insulation, for example between the outer surface of a handle, knob, grip and its shaft – unless the shaft is earthed (unless using approved knobs, handles etc. ensure that the supplementary insulation will survive ball impact and other mechanical tests – a failure under ball impact would invalidate a non-compliance within this standard).

In Class II equipment only that has a metal surface where the power cord enters the body of the equipment, supplementary insulation must be provided to the surface of the flexible supply cord (e.g. a grommet with an approved flammable rating).

Between an ELV circuit and an unearthed conductive part that would otherwise be operator accessible (ELV has only basic insulation from hazardous voltage – supplementary insulation here provides two levels of protection).

As an element of double insulation (see note at end of basic insulation).

Double or reinforced insulation
These are generally between primary circuits and unearthed accessible conductive parts (e.g. any piece of unearthed metal work which the operator can touch).

A floating SELV circuit (the SELV circuit will provide two levels of protection to match that of double or reinforced insulation).

Telephone network voltage (TNV) circuit. Note that the requirements for TNV circuitry and SELV circuitry are significantly different.

a single level of protection. When basic and supplementary insulations are combined, there are two levels of protection – hence making operator access to equipment protected by double insulation acceptable.

Reinforced insulation provides an equivalent level of protection to double insulation.

Examples of situations where these insulations may be used are:

OPERATIONAL INSULATION:

Between parts of different potential (e.g. between adjacent turns of a transformer, solenoid or motor winding – here enamel insulation provides operational insulation and allows the wound component to function correctly).

Between an ELV circuit or an SELV circuit and an earthed conductive part (e.g. low voltage ribbon and data cables and earthed metal work, or an SELV circuit and unearthed conductive parts – this latter case is allowable because a failure of insulation will expose the operator only to SELV voltages, and these do not create a hazard within the meaning of this standard).

BASIC INSULATION:

Between a part of hazardous voltage and an earthed conductive part (e.g. it is acceptable for the brown inner core of the mains cable to be placed in contact with an earthed metal part: it is *not* acceptable for a brown, or blue, conductor from a mains cable to be placed in contact with unearthed conductive parts which are operator contactable). (Note to US designers: UL1950 includes a D3 deviation that allows contact between primary insulation and exposed unearthed parts – this is not acceptable in Europe.)

Between a part that hazardous voltage and an SELV circuit which relies on being earthed for its integrity (note that this method requires special testing and that earthing of an SELV circuit is not acceptable in Denmark).

PERSONAL MESSAGE:

My personal preference is to design using method 1 or method 2. This is because I find them easier and they reduce the amount of compliance testing necessary to control during manufacture – also I feel more comfortable with these methods.

EN 60950 *Standard* *reference*	*Synopsis*
2.2.7	Deals with determining the working voltage (note that care must be taken with respect to bands of measuring equipment and signals imposed on the circuit elements involved).

When completing this section remember that creepage distance and clearance distance are treated differently under certain conditions.

Non-repetitive transients due to lightning and other atmospheric disturbances can be disregarded (these are taken into account elsewhere in the standard). The voltage of ELV or SELV circuitry is considered to be zero calculating clearance distances. However, the voltage must be taken into account before determining creepage distances.

Where a DC value is used the peak value of any superimposed ripple must be included.

The nominal value of the mains supply voltage is to be used (this would be based on the rating label).

Any unearthed accessible conductive part is assumed to be earthed and at zero potential.

Where a transformer winding or other power source is floating it should be assumed to be earthed at the point at which the highest working voltage would be obtained (e.g. if two SELV supplies of 0 and +24 V, and a floating 12 V supply are available then the two following cases must be considered: 0 V, +24 V and +36 V considering the positive half of the 12 V supply with respect to ground and other circuit elements; and −12 volts, 0 volts and +24 volts and considering the insulation between the −12 and +24 volt circuitry.

When considering insulation between windings on transformers take into account external connections that can be made when determining the highest voltage that can be generated between them.

When considering the voltage between a transformer winding and another part the highest voltage between that winding and part shall be used.

When calculating the working voltage across double (and multiple) layers of insulation each insulation element shall be considered (in turn) to be short-circuited.

(NB. This condition must not be confused with abnormal tests and considerations where single fault conditions may include leads becoming detached and touching unearthed metal parts − under these considerations secondary effects must be considered with respect to operator safety.)

If the windings configuration is not defined within the equipment and the respective outputs are floating then worst case considerations must be made.

WARNING:
Consider a bobbin with a start and finish on a single winding. It may appear reasonable to wire the inner start of the winding to live and the outer finish windings to neutral. This would appear to minimize creepage and clearance distances between the finish windings and other metal parts.

IMPORTANT NOTICE:
Phase and neutral conductors must be treated identically.

In the event of a single failure in either basic or supplementary insulation the remaining insulation system must be capable of withstanding the applied voltage.

EN 60950 Standard reference	Synopsis
2.3	SAFETY EXTRA LOW VOLTAGE CIRCUITS
2.3.1	Is the general requirement that all SELV circuits must be safe to touch under normal operating conditions and under single fault conditions.
2.3.2	This is the general requirement for SELV circuitry that any accessible voltage between two points including earth shall not exceed 42.4 V peak or 60 V d.c. under normal operating conditions.
2.3.3	This section allows the SELV 42.4 V peak and 60 V d.c. limits to be extended up to 71 V peak or 120 V d.c. for a period of up to and not exceeding 200 ms (milliseconds).
2.3.4	Method 1 establishes SELV by separation from hazardous circuitry by double or reinforced insulation. The recommended methods are:

Designing compliant SELV power supplies is a very complex task, requires considerable expertise in both design and testing – and is not to be taken lightly. In low volume and medium volume production it is often cost effective to purchase SELV power supplies which have appropriate safety approvals in place. It is not possible to overemphasize the advantages of using approved SELV power supplies: many man-weeks of design effort can easily be wasted in attempting this task. If it is absolutely necessary to design a custom SELV power supply then ensure that there the programme allows adequate contingency for redesigns and resubmission to a test laboratory.

If additional de-coupling capacitors are added to the SELV power lines ensure that the 0.1 µF limit is not exceeded.

If high current or high voltage circuits are generated from SELV ensure that under any single fault and component failure that hazardous voltages or currents cannot be injected into the SELV circuits. Check that a single failure of basic insulation (in internal wiring or across any component), or a failure of any single component (including any subsequent component failures) cannot create conditions that exceed SELV limits.

Remember that single failures will include failure of any single joint and may allow individual equipment wires to become detached and contact SELV or hazardous circuitry.

This is not the only requirement for SELV; there are also current and energy limitations.

Remember that: If there is a 0 and +50V a 0 and –50V output then the 60dc limit is exceeded.

Remember that if there is a 0 and +50 V a floating 0 and +50 V output then the 60 d.c. limit is also exceeded. This is because the two outputs can be combined to exceed the SELV voltage limit.

It is permitted to exceed the voltage limitations under the circumstances defined in section 2.3.3.

This is to allow the fault currents generated by insulation or component failures to activate overcurrent devices and settle to acceptable levels.

This section described three methods which are acceptable for designing SELV circuitry (power supplies). If the product is intended for sale in Denmark then only two of these methods may be followed (methods 1 and 2). We shall therefore only consider methods 1 and 2 in the following text.

These will provide a physical insulation which must meet creepage and clearance distances specified in the following sections. Physical barriers must take account of flammability requirements, routing must avoid sharp edges and provide either clearance or adequate insulation from hazardous

EN 60950 *Standard* *reference*	*Synopsis*
	• Provide separation by barrier routing or fixing • Provide insulation for all adjacent wiring involved that is rated for the highest working voltage present • Provide an additional layer of insulation over either SELV or hazardous circuits. • Use any other means providing equivalent insulation.
2.3.5	Method 2 where parts of SELV circuits are separated from hazardous voltages by an earthed screen or other earthed conductive parts.
2.3.6	Method 3 is not acceptable in France.
2.3.7	This section of the standard is void.

voltages; fixings in this application will be safety critical and therefore should not rely upon self-adhesive tape or other similar methods which may fail at elevated temperatures.

If hazardous voltages are present on basic insulation (e.g. 300 V) then SELV wiring must be protected by similar insulation. If the basic insulation protecting the hazardous voltage fails there is another level of basic insulation protecting the SELV wiring from hazardous voltage.

NB. Remember to consider the highest available potential difference during normal operational use; if this exceeds the voltage rating of either wire then the design will be non-compliant. Provide insulation on either the wiring of the SELV circuit or that of the other circuits that meet the insulation requirements for supplementary or reinforced insulation as appropriate for the highest working voltage present.

Where SELV wiring may contact bare hazardous parts or hazardous parts protected only by operational insulation the SELV wiring must be protected by at least reinforced insulation. Conversely if wiring carrying hazardous voltages may contact bare SELV parts or SELV parts covered only by operational insulation then the hazardous voltages must be protected by at least reinforced insulation (when hazardous voltages are routed through equipment consider using normal power cord: this will provide basic plus supplementary insulation and will reduce investigations necessary to prove compliance, fault testing, earthing requirements and, provided the terminations at either end remain short, will reduce the risk of broken wires – single fault conditions – creating a hazardous voltage).

Where hazardous voltage and SELV are locally in close proximity supplementary insulation may be provided over basic insulation, or basic insulation may be applied over operational insulation or bare parts, to provide the necessary isolation. (NB. Ensure that wires cannot become loose and compromise safety.)

This section allows the designer additional freedom in providing the necessary insulation between hazardous and SELV circuits.

The hazardous voltage and earthed screen are separated by at least basic insulation. The function of the earthed screen is to protect SELV circuitry from any failure of insulation. The commonest place to see this form of protection is on the underside of power supplies where an earthed screen may be run to provide a barrier between hazardous and SELV.

Make sure that the earthed screen can carry the fault current reliably; if the screen melts or vaporizes it will have little value as a protective device.

This method is not described because it is not accepted in France.

EN 60950 Standard reference	Synopsis
2.3.8	Deals with four general construction requirements. 'Ring-tongue' and similar terminations shall be prevented from pivoting, which would reduce creepage distances and clearance between SELV circuits and parts at hazardous voltage. In multiway plugs and sockets and wherever shorting could otherwise occur a physical means should be provided to prevent contact between SELV circuits and parts at hazardous voltage due to loosening of the termination or breaking of a wire at termination. Uninsulated parts at hazardous voltage should be located or guarded so as to avoid accidental shorting to SELV, for example by tools or test probes used by service personnel. SELV circuits shall not use connectors compatible with those specified by IEC 83 or IEC 380.
2.3.9	This section covers the requirements when connecting SELV circuits to circuits of other categories. These connections may be either signal level, or taking power for SELV circuitry from the other circuit. The conditions that must be met are: SELV circuits must not be conductively connected to any primary circuit The SELV circuit must meet the SELV limits under normal operating conditions while the connection is made. Where the SELV circuit obtains its supply by connection to a secondary circuit and that secondary is separated from primary or hazardous voltages by double insulation (method 1 in the previous section) or by the use of an earth conductive screen (method 2) and there is a minimum of basic insulation between the primary or hazardous voltage and the earth conductive screen, then the SELV circuit shall be considered as separated from the primary or hazardous circuit.
2.3.10	Compliance with sections 2.3.1 to 2.3.9 is checked by inspection and appropriate tests.

Guidance notes

Many approved termination strips have barriers to prevent this from occurring, alternatively ('UL recognized') heat shrink sleeving may be used to provide additional levels of insulation, and meet flammability requirements.

Here we are considering a single failure only.

If the copper strands of an electrical wire are considered to break, it is normal to assume that the insulation covering that wire will *not* break.

Hints. Terminations that grip the wire and also the outer insulation should not normally need to be considered.

Where a wire breaking would allow a lead freedom to move then the lead should be swept over all possible parts of circuitry to ensure that such a single fault will not compromise SELV circuitry.

The use of tie wraps or sleeving will reduce the distance that broken leads can move through.

In multiway plugs ensure that there is adequate creepage and clearance distances particularly where the wire attaches to the termination.

Do not use mains connectors for SELV applications – the reason for this is obvious.

(Thirty years ago I knew someone who installed mains plugs and sockets for loudspeaker systems in a hall. One day his son plugged a 15 Ω loudspeaker into the 230 V mains supply!).

Note also that the use of 4 mm 'banana plugs' has been expressly prohibited by EN 60065 – they fit into the European mains socket.

This includes neutral, because where mains connectors are not polarized, neutral and live could be transposed. Conductive connections would be either resistive or inductive: capacitive coupling would be acceptable provided that the double insulation criteria are met.

This is logical, since it may be possible to either raise the voltage or the supply currents beyond the SELV limits.

There is very little difference between these requirements and those general requirements for SELV circuitry.

For connection with TNV circuits please check section 6.

Take circuit diagrams and printed circuit board artwork and highlight primary circuitry with red pens and SELV with green. This will identify

EN 60950 Standard reference	Synopsis
2.4	LIMITED CURRENT CIRCUITS
2.4.1	Requires that limited current circuits do not exceed specified limits under normal operating conditions or following a failure of basic insulation, or any single component failure. The segregation of limited current circuits from other circuits is described in sections 2.3 and 2.4.6.
2.4.2 to 2.4.5	These sections define the absolute maximum limits permitted for limited current circuits.
2.5	PROVISIONS FOR PROTECTIVE EARTHING
2.5.1	Accessible conductive parts of Class I equipment that could become hazardous in the event of a single insulation failure must be reliably connected to a protective earthing terminal within the equipment. In service access areas where conductive parts may assume hazardous voltage in the event of insulation failure, it is permitted to provide a suitable warning label indicating that parts are not earthed and should be

and highlight areas where either reinforced or basic plus earthed screen or basic plus supplementary insulation must be in place.

When performing electric strength measurements or tests on the equipment, first perform tests for basic insulation.

Next disconnect all circuitry and circuit elements which are separated from ground by only basic insulation.

Finally, perform reinforced tests upon the remaining circuit elements.

Limited current circuits may be implemented using resistors or capacitors.

Example. The theoretical series resistance for a 240 V 50 Hz mains supply is a series resistance of 350 kΩ.

It is not acceptable to use one single resistor of, say, 390 kΩ because in the event of a single component failure there would be nothing to limit the current.

Similarly, in the event of a single failure the use of two resistors of 180 kΩ each in series would fail to meet the limited current circuit, because the failure of either resistor would increase the current above the 0.7 mA limit.

A limited current circuit providing 0.4 mA could be implemented from the 240 V mains using three 180 kΩ resistors in series. In the event of a single component failure the current would remain below the 0.7 mA limit.

NB. The resistors must be selected so that their power and current ratings are adequate should only two be in circuit at 240 V. The creepage and clearance paths across these components must be carefully investigated for compliance given any single component failure.

Because limited current circuits may be touched by an operator they must follow the same segregation requirements as SELV circuits.

This is the maximum permitted after a single component failure (including subsequent breakdown or failures) or failures of basic insulation.

There are additional requirements when connecting to TNV (telephone network voltage) circuits in sections 6.3.2 and 6.3.3.

Accessible means accessible by an operator.

This includes using the test finger, after using the steel ball and other tests.

A single insulation failure means a failure in basic or supplementary insulation only: it does not require a failure of reinforced insulation to be considered.

Reliably connected means connected in a positive manner that is unlikely to fail.

EN 60950 Standard reference	Synopsis
	checked before touching.
2.5.2	States that Class II equipment shall have no provision for protective earthing except where continuity of protective earthing is provided to other circuits. If Class II equipment has a functional earth, this must be separated from Class II parts by double or reinforced insulation.
2.5.3	Protective earthing conductors shall not contain switches or fuses.
2.5.4	Requires that systems comprising Class I and Class II equipment shall be interconnected so that earthing of all Class I equipment is assured, regardless of the arrangements of the equipment in the system.
2.5.5	Describes the earthing conductors that are permitted.
2.5.6	Requires earth connections to be designed so that disconnecting a protective earth from one assembly will not break the protective earth connection to other assemblies, unless hazardous voltages are first removed.

Guidance notes

The standard suggests motor frames and electronic chassis as examples of items that should be marked. When considering the extent and detail to which warnings should be given, remember that we can assume service personnel will be competent in dealing with obvious hazards: pay special attention to potential hazards which may not be obvious, e.g. large heat sinks or busbars which may appear to be at earth potential.

NB. Remember that an air gap providing reinforced insulation between hazardous voltage and an accessible conductive part of an enclosure (floor standing equipment or the non-vertical top surface of desktop equipment) must provide a clearance of not less than 10 mm (Table 3 note 6 and Table 5 note 7). These clearance distances must be achieved even when the equipment is subject to the mechanical forces defined in sections 2.9.2 and 4.2.3.

Compliance with these requirements would be carried out by inspection and by the application of earth continuity tests and electric strength measurements.

This provides clarification over the definition of Class I and Class II equipment.

The functional earth may be to provide a 0V reference for signal inputs and outputs or for connecting to an EMC ground elsewhere within the system. The need to provide reinforced insulation or double insulation between this functional earth and hazardous circuits is to prevent the generation of hazardous voltages on this functional earth in the event of a single failure of basic insulation.

Safety Earth Ground should not have any means of disconnection – unless phases are first disconnected.

If the system may be configured by an operator, then all scenarios and configurations must be considered.

If the system is rack mounted and the equipments cannot be moved without the use of tools (e.g. screwdriver) then restraining interconnecting cables may eliminate the possibility of incorrect or hazardous connections.

If the equipment configuration is restricted to service personnel only, then a suitable safety warning should be given in the installation and servicing instructions.

When using ribbon cables or flexible printed circuits ensure that the conductors will carry all likely fault currents.

It is particularly important to review this in large systems. The discipline of using three core mains cable (in single phase applications) will reduce the likelihood of a non-compliance. Take particular care where modules are removed for servicing. It may be necessary to provide additional earthing points and connections for this purpose.

EN 60950 Standard reference	Synopsis
2.5.7	Requires that operator removable parts having protective earth connections shall be of a 'first make, last break' type.
2.5.8	Requires that protective earthing connections do not need to be removed during servicing other than to remove the part which they protect unless a hazardous voltage is removed from that part at the same time.
2.5.9	Requires that non–detachable power supply cords meet the size and connection requirements of section 3.3 and shall generally be suitable for the purpose intended.
2.5.10	Considers corrosion and the electrochemical reaction resulting from combinations of dissimilar metals at protective earth connections.
2.5.11	Defines the maximum allowable resistance between the protective earthing terminal and the earthing contact on the parts to be earthed.
2.6	PRIMARY POWER ISOLATION
2.6.1	Requires a disconnect device to be provided to enable servicing.

Guidance notes

Using approved connectors designed for mains and hazardous voltages will ensure this. Take particular care where using other connector types (e.g. MIL-38999).

Never run hazardous voltages and Safety Earth Ground through separate connectors which are operator accessible (the earth could be unplugged).

When using an earthing stud as a star point for protective earthing, consider carefully the order in which protective earths are assembled.

Generally, the first protective earth (e.g. the bottom one) will be from the mains inlet: this should be locked with nuts and suitable washers before other protective earths are made off at the top of this stud. This first protective earth connection should never need to be disturbed.

Try to avoid disturbing other earth connections to this stud. If this is unavoidable during servicing include a warning to remove hazardous voltages to the equipment until all protective earth connections have been restored.

Non-detachable power cords are permanently connected to the equipment and have a plug on the other end (if the plug is removed and the non-detachable lead is connected to building wiring it becomes classified as permanently connected – see later sections for permanently connected equipment). Other considerations will be the flammability, mechanical strength and clamping characteristics of grommets used to retain the mains cord, and the length of cabling provided (for conformance with local or national wiring regulations).

It is essential that the electrochemical potentials between surfaces do not exceed 0.6 V. Ensure that engineering and design staff understand the need to control these components, and that changes are not introduced during manufacturing without proper assessment.

For equipment with detachable power cords measure between the earth pin of the inlet socket on the equipment and operator contactable metal work which is protected from hazardous voltage by only basic insulation or which could be contacted by hazardous voltage in the event of single failure (e.g. fracture of a wire, or failure of a single connector).

Record the resistance measurements made and the positions of each measuring probe. This will be vital if it is necessary to repeat these measurements at some future date.

EN 60950 Standard reference	Synopsis
2.6.2	Requires that the disconnect device has a contact separation of at least 3 mm and that it should be connected as closely as practicable to the incoming supply.
2.6.3	Requires that the disconnect device should be included either within the equipment or that installation instructions (see section 1.7.2) define the need and characteristics of the disconnect device.
2.6.4	Requires that parts on the supply side of the disconnect device within the equipment, which remain energized when the disconnect device is switched off, must be guarded to prevent accidental contact by service personnel.
2.6.5	Prohibits isolating switches to be fitted within the power cord.
2.6.6	Single phase equipment connected to a polarized supply may have a single pole disconnect device in the live. User and surface instructions must state that an additional two-pole disconnect device must be fitted in the building installation where non-polarized supplies and connectors are used. If a two-pole disconnect device is fitted to single phase equipment there is no need to include any warnings. The standard gives examples of applications where a two-pole disconnect device would be required.
2.6.7	For three-phase equipment, the disconnect device is required to disconnect all phase conductors simultaneously. However, where the neutral connection is made to an IT power system, the disconnect device must be a four-pole device and include all phases and neutral conductor simultaneously. Any disconnect device that interrupts the neutral conductor must simultaneously interrupt all phase conductors.
2.6.8	Requires switches which are used for disconnect devices to have their on and off positions marked in accordance with section 1.7.8.

Guidance notes

Note that a contact separation of 2.98 mm is a failure!

If the mains switch does not provide sufficient isolation it is permissible to use circuit breakers, isolation switches, appliance connectors or the mains plug to provide the necessary isolation. It is customary to include a servicing instruction 'Remove power cord before servicing' – this warning should be repeated in servicing instructions.

Remember that written instructions form part of the product – if we fail to give vital information then the product is non-compliant.

Consider the words 'accidental contact' – it is acceptable for some hazardous live parts to be exposed (e.g. the screws in terminal blocks, which would require a tool for access).

Some standards allow isolating switches to be fitted in the power cord.

This particularly covers equipment designed in the UK and supplied on to the mainland continent where plugs may be inverted so that live and neutral are transposed. My personal preference is to fit a two-pole disconnect device for any single phase equipment and include a fuse in the live and a separate fuse in the neutral.

These examples should be included in installation and servicing instructions.

If a four-pole device is not provided with the equipment, then the installation instructions must specify the need for a four-pole device (that operates simultaneously on all poles) as part of the building installation.

Simultaneous is usually taken to mean not requiring a second operation.

The 'O' and '|' symbols must be used: for a disconnect device. Some agencies accept that if the power cord is the disconnect device then the 'O' and '|' symbols are not mandatory on the mains switch.

EN 60950 Standard reference	Synopsis
2.6.9	Requires instructions to be given if the power cord is to be used as a disconnect device.
2.6.10	Requires that the protective earthing connection in supply plugs and couplers shall be the first connections to make and the last connections to break with respect to supply and hazardous voltages.
2.6.11	Considers the interconnection of hazardous voltages and hazardous energy levels between units.
2.6.12	For units which receive power from more than one source, there must be a prominent marking at each disconnect device, giving adequate instructions for the removal of all power from the unit.
2.6.13	Covers compliance.
2.7	OVERCURRENT AND EARTH FAULT PROTECTION IN PRIMARY CIRCUITS
2.7.1	Basic requirements. There are three basic requirements within this section, for clarity we will take them in reverse order: (c) Where a pluggable equipment type A (domestic plug is used the building wiring may be considered to provide protection in accordance with the rating of the wall socket outlet. (b) Components in series with the mains inlet may rely on protective devices in the installation to provide short-circuit and earth fault protection. (a) Any other protective device necessary to prevent hazards during single fault abnormal testing (section 5.4) should be included as an integral part of the equipment.
2.7.2	Void

Guidance notes

My personal preference is to comply with, or preferably exceed, the requirements of the standard – if we design to the minimum requirements we do not allow any room for error.

By using a suitable disconnect device (e.g. a two-pole switch with 3mm clearance), the need for these instructions is avoided.

Use of approved connectors for these applications should ensure this requirement is met.

This requires either separate disconnect devices to be fitted, or that parts should be guarded to reduce hazards to service personnel and that appropriate markings, warning labels and service instructions shall be provided with the unit.

Suitable wording for a unit with four power cords may be 'Disconnect four power cords before servicing this unit'.

If the units are rack mounted and the power cords are part of the internal fixed loom, then it is appropriate to number each power cord 1 to 4.

Although not specifically covered within this standard, the use of UPSs (uninterruptible power supplies) would require clear marking and servicing instructions to eliminate hazards.

For example, a 13 A socket and plug will provide fault protection for wiring and components rated at 13 A, a 16 A socket and plug will provide overload and short-circuit protection for components and wiring rated at 16 A.

Ensure that components in this circuit are adequately rated (e.g. the power supply cord, RFI filter and switch must be rated at the same current as the earth fault protection in the installation).

Example. If building wiring provides earth fault protection at 16 A and abnormal testing shows a need to provide additional fault protection at 2 A then the 2 A protective device must be included within the equipment. The protective device does not need to be operator or service accessible to comply with this standard.

EN 60950 Standard reference	*Synopsis*

2.7.3	Short-circuit protection. This deals with the breaking capacity of protective devices under short-circuit conditions.
	   Pluggable equipment type A may use building installations to provide short-circuit protection.
	   With permanently connected equipment or pluggable equipment type B, it is permitted to use short-circuit backup protection in the building installation.

2.7.4	Number and location of protective devices. The general requirement is to have sufficient protective systems or devices located so as to detect and interrupt excessive current flowing in any possible fault current path. The examples given are phase to phase, phase to neutral and, in the case of Class I only, phase to protective earthing. In the case of 3 phase equipment any protective device that interrupts the neutral conductor must also interrupt all other supply conductors. The use of single pole protective devices in 3 phase applications is absolutely and expressly forbidden. Table 1 illustrates the minimum numbers of fuses or circuit breakers that are required to detect earth faults and overcurrent faults in various configurations of equipment. It also identifies where they need to be placed.
2.7.5	Allows several protection devices to be combined in one component.
2.7.6	Requires warnings to be given to service personnel where they may be exposed to hazards following the operation of a protection device.
2.8	SAFETY INTERLOCKS
2.8.1	Requires safety interlocks to be provided where there is operator access to areas which would otherwise present a hazard within the meaning of this standard.
2.8.2	Requires safety interlocks to be designed so that the hazard is removed before there can be operator contact

Guidance notes

Under short-circuit conditions fault currents of many thousands of amps may flow. These high currents can exceed the rupturing current capacity of protective devices such as circuit breakers and certain types of fuse.

This means domestic plugs and sockets, provided that the mains cables and components are adequately rated for the circuit.

Pluggable type B refers to industrial connectors (e.g. IEC 309 type). However, the requirement and the characteristics of the additional short-circuit backup protection must be given in installation and servicing notes.

The suitability for short-circuit protection will be tested as part of abnormal tests in section 5.4.

Do consider additional applications and marketplaces before commencing the design of equipment. If you manufacture for the UK or US market, consider the need for two protection devices; to detect earth faults in both the live and neutral. This will be necessary if you wish to supply to the whole of Europe.

A multipole circuit breaker (with 3 mm contact clearance distance) can provide overload and short-circuit protection and act as the disconnect device.

The examples given are where protection devices are fitted to the neutral of single phase Class I equipment and other instances where the operation of a protection device is not intended to remove all hazardous voltages from the equipment.

Suggested wording is 'Caution double-pole/neutral fusing'.

You must consider the operator to be oblivious of hazards and therefore provide suitable protection. Warning notices are inadequate for extreme hazards (e.g. live parts).

Remember that bleed-resistors used to discharge capacitors may fail open-circuit. It is therefore prudent to consider this failure mode when

EN 60950 Standard reference	*Synopsis*
	with it. This is defined in detail in that section. In particular, it defines the maximum time to reduce the voltages and energy levels to acceptable limits.
2.8.3	Defines the criteria for inadvertent operation of the safety interlock.
2.8.4	Provides two alternative methods of evaluating the suitability of an interlock system. The first is to analyse probable failure modes to ensure that they would not create a hazard. The second merely considers the probability of failure within the normal life of the equipment, given that any failure will not cause an extreme hazard.
2.8.5	Deals with the need for service personnel to override safety interlocks.
2.8.6	Requires the mechanical interlock switch to meet certain clearance distances defined in section 2.8.6.1 or to pass tests in sections 2.8.6.2 and 2.8.6.3.
2.8.7	Requires mechanical interlock systems to have built-in over-travel.
2.9	CREEPAGE CLEARANCE DISTANCES AND DISTANCES THROUGH INSULATION
2.10	CONNECTION TO OTHER EQUIPMENT

Guidance notes

calculating time constants and the number of resistors that should be used.

Verify the suitability of these safety interlock requirements in respect of local health and safety requirements and the potential hazards that may occur (e.g. the definition and requirements for safety interlocks within this standard may not be acceptable for guarding certain types of hazardous machinery).

Consider product liability implications of such an analysis and in particular, when using the second approach, that however improbable, there will be a finite possibility that a hazard will occur during the life of one piece of equipment.

Remember that it is not the company that accepts that risk: it is your personal liability.

Ensure that components selected in safety interlock applications are suitable for those applications and have an appropriate safety approval. (E.g. Microswitches may weld and fail to operate electrically although the mechanism will function normally. Positive displacement switches – where the operating mechanism moves the contacts directly – are less prone to this failure.)

These are defined clearly in the text of the standard although the exact definition of 'extreme hazard' is omitted.

Microswitches are not suitable as safety interlocks, we need a 'positive displacement' type of switch. If the hazard is extreme then we need to check if the components and circuit need to be assessed under Annex D of the Machinery Directive.

This is to reduce the risk of the interlock system becoming damaged when it is exercised beyond its normal operating position.

These requirements are specific to detailed component and printed circuit board layout. Printed circuit manufacturers will have software for making these calculations based upon the requirements of EN 60950, the installation category, pollution degree and level of isolation required.

Components connected to primary or hazardous voltages will be selected with appropriate safety approvals in place and when used within their rating will comply with these requirements.

Please refer to Chapter 8 which provides worked examples of these calculations.

EN 60950 Standard reference	Synopsis
2.10.1	Requires that where equipment is intended to be electrically connected to other equipment, requirements for SELV and TNV circuits in section 2.3 and section 6.0 shall be maintained. The notes to this section support the use of a single interconnecting cable carrying a mixture of SELV, limited current TNV, ELV and hazardous voltage.
2.10.2	Restricts the use of ELV interconnection circuits.
2.10.3	Restricts and defines ELV inter-connection circuits to complementary (add on) equipment.
2.11	LIMITED POWER SOURCE. The definition includes references to the use of an isolating transformer and the use of non-adjustable, non-automatically reset electromechanical devices for overcurrent protection.
3.0	WIRING, CONNECTIONS AND SUPPLY
3.1	GENERAL
3.1.1	This covers rating of wires and protection.
3.1.2	Deals with wire ways and routing of cables.

Guidance notes

Connecting SELV circuits with SELV circuits or TNV circuits with other TNV circuits will usually achieve this requirement. However, take caution where circuits are not referenced to earth and combined voltages may exceed SELV limits. Alternatively facing circuits in parallel may exceed the allowable current power limits for SELV.

When considering a single failure of basic insulation between these wires, consider also the failure of a single joint within a plug or a socket.

Limits interconnection circuits to SELV, or TNV, or hazard voltage except where combinations are permitted by 2.10.3.

ELV provides only one level of protection from hazardous voltage, it should therefore be assessed in its intended and 'likely' applications and during connection.

Isolating transformers should comply with Annex C2 – Safety Isolating Transformers – with respect to their construction and electric strength requirements.

Current limiting overcurrent protection is typically achieved using fuses. Note that at low voltages (below 10 V) some fuses will fail to rupture, considerable evaluation may be necessary at these extreme voltages.

Inherently limited power supplies may be designed using transformers whose losses are such that the limits defined in Table 8 (of EN 60950) cannot be exceeded. If rechargeable batteries are used to provide standby power, then fuses should be included to limit power available as defined in Table 9.

In the first edition of EN 60950 Table 9 output current short-circuit included a typographical error which has since been corrected ($1000/U$ was $1000\ U$). This caused great confusion, some of which still remains to this day.

All internal and external wires, cable and locks must be adequately rated and protected from overload and short-circuit.

In high power circuits check the wire temperature – if the wiring heats it may be cheaper to improve cooling than to increase the section of copper.

The first paragraph refers to holes in metal and the use of bushings. Where wires containing hazardous voltage pass through metal in a Class II equipment, refer to the special requirements for grommets mentioned in previous sections. Note also flammability requirements.

EN 60950 Standard reference	Synopsis
3.1.3	Deals with routing and supporting of internal wiring.
3.1.4	Relates to creepage and clearance distances of uninsulated conductors.
3.1.5	Deals with the suitable working voltage for individual conductors.
3.1.6	Restricts the use of yellow/green wires for protective earth connections only.
3.1.7	Deals with bead and ceramic insulators.
3.1.8	Requires two complete threads engagement of screws used for electrical connections. Prohibits the use of screws made of insulating materials (e.g. nylon) where used for insulating purposes and their replacement by a metal screw could impair supplementary or reinforced insulation.
3.1.9	Electrical connections shall be so designed that contact pressure is not transmitted through insulating material unless there is sufficient in the metallic parts to compensate for any shrinkage or distortion of the insulating materials.
3.1.10	Deals with multistranded conductors and contact pressure, in particular clamping screws.
3.1.11	Deals with the use of spaced-thread and thread-cutting screws for current carrying and protective earthing applications.

Guidance notes

Do not flex or strain wires and terminations; do not cut or damage insulation.

Be careful where Safety Earth Ground connections are to moving parts (e.g. a door or lid). It is often worth considering if a Safety Earth Ground connection is *really* necessary or how the need can be avoided.

We assess if the creepage and clearance can be reduced.

Specifies the need for testing.

Functional earthing should be green or any other colour.

If you have incorrectly identified functional grounds as protective earth connections and the product is certified by an independent agency, remember that all yellow/green wires must be present until that agency has agreed to their removal. (It may affect safety marking of equipment if modifications are made.)

This is specific to high temperature items.

The requirement for a minimum of two complete threads engagement for screw threads is good engineering practice.

Take particular care when fitting primary earth points studs onto plastic mouldings and parts. Do not sandwich the electrical connections with the plastic mouldings or parts: they are likely to fail as the plastic ages or softens.

It used to be common practice to dip multistrand conductors into soft solder as it made them easier to fit into connectors and connect to blocks. Experience showed that the solder was subject to cold flow and released contact pressure which gave rise to hazard and equipment failures.

These screws are permitted for earth continuity provided that they are not disturbed after connection has been made and that at least two screws are used in each connection. Remember to consider the electrochemical potentials in Annex J of the standard.

Remember that a screw used for earthing may not also have any other function (e.g. if two screws are used for earthing then neither of these

EN 60950 Standard reference	Synopsis
3.2	CONNECTION TO PRIMARY POWER
3.2.1	Defines the method of connection.
3.2.2	Details the requirement for permanently connected equipment.
3.2.3	Defines the requirements for appliance inlets.
3.2.4	Defines the suitable power cords.
3.2.5	Covers very detailed requirements and test requirements for non–detachable power supply cords.

designated screws may have the mechanical function of holding parts together).

The standard prohibits operator contact with bare parts at ELV or hazardous voltage. This is to ensure that there can be no operator contact while plugs and sockets are inserted. Use test finger for this purpose.

Note the minimum sizes for cable and conduit entry – Table 10 of the standard.

Ensure that there is sufficient clearance for conduit nuts etc. and that creepage and clearance distances are not compromised.

For Class I equipment ensure that the earthing terminal of the inlet socket connects to the primary protective earthing point of the equipment – ideally this should be made off with locking nuts so that it is never disturbed when other protective earthing terminals are placed on top of the stud.

Note that power cables identified with the '<HAR>' are harmonized power cables which will fit all plugs within the European Union.

This is to ensure that the power supply cord will not create a hazard if it is stressed in normal use. Typical considerations are if the cord is pulled and slips through the anchor points the last connection that must break must be the Safety Earth Ground. Always ensure that there is slack in the earth wire in the equipment and also in a re-wireable plug.

Similarly, if the cord can be pushed into the equipment in such a way that the power cord itself, or internal parts of the equipment, could create a hazard or be displaced.

It should not be possible to rotate the cord if that would apply a stress to electrical connections.

Other considerations are if the cord is pulled to an extent that the insulation is damaged, the use of supplementary insulation within the anchorage device or cable clamp may be required.

When selecting cable clamps it is often worth spending a little more on a good quality component that exceeds electrical and flammability requirements (e.g. UL94-5V).

Consider also how a cord will be replaced: will the service engineer understand any limitations on clamping or other ways in which they could foreseeably introduce a hazard into the equipment. Check wording of standard for details of tests and pass criteria.

EN 60950 Standard reference	Synopsis
3.2.6	Deals with the exposure of power supply cords to sharp edges within the equipment and defines the requirements for inlet bushes, cord guards etc.
3.2.7	Requires a cord guard to be fitted where non-detachable power supply cords are used on hand-held equipment, or equipment which is intended to be moved while in operation.
3.2.8	Requires adequate space to be provided so that connections for non-detachable power supply cords or permanently connected equipment can be made reliably, to allow checking of the connections, to ensure connections are easy and should they become loose will not contact any accessible conductive parts.
3.3	WIRING TERMINALS FOR EXTERNAL PRIMARY POWER SUPPLY CONDUCTORS
3.3.1	Deals with permanently connected equipment and equipment with a non-detachable power supply cord.
3.3.2	Allows any reliable electrical and mechanical connection to be used when joining special non-detachable power supply cords to the internal wiring of an equipment, provided that permitted temperature limits are not exceeded.
3.3.3	Details the thread to be used on power supply connections.
3.3.4	Outlines the velocity for power supply cords: that two independent fixings are not considered likely to become loose at the same time. Soldered connections are not adequate unless they are

Guidance notes

Note that inlet bushings and anchorage devices must not be removable without the use of a tool and must be reliably fixed. There are special requirements for non-metallic enclosures and metal-cased Class II equipment: these and other requirements are most easily met by selecting a good quality plastic cable anchorage.

This requirement would apply only to power cords which supply hazardous voltage, current or energy to the equipment. This section also contains details of the tests required to demonstrate compliance. (Look at the one fitted to your domestic iron.)

Ensure there is adequate room for tools that may be necessary and that any stud used is adequately locked so that it cannot become loose and compromise contact pressure.

Most important of all, ensure that there is sufficient space to provide access and good visibility of the connections.

Refer to Table 13 of the standard for the largest conductors that must be accepted by the terminals.

Unless your installation instructions specify the use of flexible cords then the column marked 'other cables' must be used. If the terminals are used for primary power connection, note also Table 14 which defines the minimum acceptable diameter of studs and screw type connectors.

These requirements should not be compromised by lack of space.

Requires that connection be made by means of screws, nuts or equally effective devices – so that the cord or wiring can be made or replaced.

Soldered terminations must first be hooked or mechanically fixed together.

If sleeving is used remember the creepage distance between the cable and the internal surface of the sleeving, and allow for manufacturing/assembly tolerances. Check that manufacturing instructions and processes ensure that there are no sharp edges that may cut the sleeving.

Confirm compliance as detailed in the standard.

When mounting conductors to a printed circuit board, two acceptable methods are: applying a crimped connector to the wire and soldering the connector to the printed circuit board; alternatively pretinning the wire passing the pretinned end through a suitable hole in the printed circuit

EN 60950 Standard reference	Synopsis
	held in place near the soldered joint.
	Conductors which are connected to terminals by other means are not considered adequately fixed unless there is an additional fixing near the terminal.
3.3.5	Refer to Table 13 of the standard for the nominal cross-sectional area of conductors.
3.3.6	Defines the minimum sizes of terminals and requires stud terminals to be provided with washers.
3.3.7	Requires terminals to clamp the conductor between metal surfaces with sufficient contact pressure and without damage to the conductor. The terminations may be clamped or fixed so that: • the terminal itself does not work loose; • internal wiring is not stressed; • creepage and clearance distances are not compromised.
3.3.8	'For ordinary non–detachable power supply cords, each terminal shall be located in proximity to its corresponding terminal or terminals of different potential and to the protective earth terminal, if any'.
3.3.9	Deals with stray strands from a flexible conductor.
4.0	PHYSICAL REQUIREMENTS

board, bending the wire at 90° and soldering.

When joining two wires together by soldering the two pretinned ends of the wires should be hooked together (e.g. bent 180° so that the two wires interlock before they are soldered). See Figure 7.6.

Consider a crimp connection to a wire. We must assume that one fixing may fail, therefore a single crimp to the conductor will be inadequate for reliable termination.

We must therefore use a crimp that will fasten to the conductor and a second crimp which will secure the insulator of the wire.

When using single spade or other similar connectors remember again that two independent fixings must be used to achieve a reliable connection.

For earth connectors, and where similar reliable connections are required, choose a connector which provides a positive indent or lock, so that it cannot be pulled off.

In large installations, it may be prudent to specify a torque to be applied to the connectors to avoid damaging them. If the conductors may slip out when screws or nuts are tightened ensure that service instructions define the correct method of assembly. As with all connectors and terminals check that the smallest and the largest cross-sectional areas of conductors fit correctly. If the largest section specified in Table 13 will not fit then larger connectors will be required. If the smallest conductor specified in Table 13 will not fit then servicing instructions must specify the size of conductors that must be used.

Input terminals should be grouped together with any protective earth terminals specific to them.

Input terminals should be located together with their respective earths.

Single phase connections should be grouped together with their respective earths.

Three-phase connections should be grouped together with their respective earths.

Refer to text in standard which details the considerations and tests required.

NB. We are forbidden to use soft solder to hold strands together before they are clamped.

EN 60950 Standard reference	Synopsis
4.1	STABILITY AND MECHANICAL HAZARDS
4.1.1	Requires that equipment should not present a physical hazard to operators or service personnel under conditions of normal use.
4.1.2	This section generally requires that hazardous parts should be guarded. Requires that the operator should be prevented from accessing hazardous moving parts. Service personnel shall be protected from unexpected and unintentional contact with hazardous moving parts.
4.1.3	This allows operator access to hazardous moving parts under certain, well-defined, conditions.
4.1.4	Requires edges and corners to be round or smooth to avoid operator injury. Sharp or serrated edges are permissible where these are required for the correct functioning of the equipment.
4.1.5	Requires a warning label on high pressure lamps.
4.2	MECHANICAL STRENGTH AND STRESS RELIEF
4.2.1	General This section defines the acceptance criteria and test methods to ensure that enclosures have adequate strength and will not cause a hazard during normal operation or use. These requirements do not apply to internal fire

Guidance notes

If the equipment is transported in special packing or on a pallet then a servicing stability (e.g. 10°) test should be conducted on the transported package, the equipment and at interim stages of unpacking: this will identify any operations where service personnel require special warning or instructions. When applying forces to the equipment the equipment must not be allowed to slide.

Please refer to standard for details of tests.

Read this section with 4.1.3.

Unless covered by 4.1.3, put hazardous moving parts in service access areas and include servicing instructions that will remove power and allow these parts to cease moving.

Alternatively, provide mechanical or electrical safety interlocks that will remove the hazard before an operator can gain access to those areas.

Hazards should be clearly visible or obvious. Special protection will be required where timing devices or automatically resetting thermal cut-outs may cause unexpected operation within the equipment. For detailed test requirements, refer to the standard.

Provided that the hazardous moving part is required for the function of the equipment and the process (e.g. removing the hazardous moving part would destroy the functionality of the equipment) and where the hazard is obvious to an operator then a warning in the local language or international signs is considered acceptable. NB. If fingers, jewellery, clothing can be drawn into the moving part then there must be a suitable means to stop the equipment.

Any interlock, emergency stop or warning label is a safety critical item and should be controlled accordingly.

It would be prudent to place a suitable warning note in the operator instructions, and perhaps close to the potential hazard.

The way to consider this is as follows:

It is acceptable that the enclosure may fail to meet these requirements provided that the operator cannot access or be exposed to a hazard.

If a window fails one of the following tests then the window should be removed and the test finger used to assess whether a hazard would be accessible.

EN 60950 Standard reference	Synopsis
	enclosures or internal barriers which form part of a fire enclosure. Nor do they apply to: handles; levers; knobs; the face of cathode-ray tubes which comply with IEC 65 which have faces exceeding 160mm; or certain transparent covers for meters; measuring devices, etc. unless parts at hazardous voltage would be accessible if the cover was removed.
4.2.2	Steady force test. The 30 N force test is intended to ensure that components (particularly plastic parts like switches) do not yield under pressure to expose hazardous live parts.
4.2.3	Steady state force. The 250 N force test is particularly applicable to metal enclosures where flexing of the external surfaces may reduce clearance distances below the 10mm limit required.
4.2.4	Steel ball test. This impact test must be conducted on enclosures that restrict operator access to hazards.
4.2.5	Defines the test methods for hand-held and direct plug-in equipment.
4.2.6	Stress relief test. This test is conducted on moulded or formed plastic enclosures at a minimum temperature of 70°C or 10° above the maximum observed temperature.
4.2.7	Deals with the compliance criteria; refer also to the use of the test pin and test finger in section 2.1.2.

Guidance notes

Similarly, if the window or section of the enclosure which failed the test was required to fulfil the functions of a fire enclosure then the equipment would fail to comply with this standard.

Exercise great care when reviewing, assessing or testing equipment to these series of tests.

Where plastic parts are involved, this test must be repeated at the relevant maximum temperature defined by this standard.

(Plastic parts become very supple at elevated temperatures and it is not uncommon for failures to occur.)

Check also plastic parts as in the above example and that sections held by glue/cement or plastic retainers do not fail under normal or elevated temperatures.

It is acceptable for the surface to break provided hazards cannot be contacted with the test finger and creepage and clearance distances are not compromised.

It is acceptable to rotate the test sample 90° so that a vertical impact can be applied rather than a side impact with pendulum.

Tongue and grooved oak flooring is not easily obtained: it may be necessary to resort to a joinery.

The standard assumes temperature rises will be based on a maximum ambient temperature of 25°C. For the purpose of this test, raise to a temperature of 10° above the maximum worst case temperature. The following two examples may clarify the temperature rise measured during testing, maximum operating (ambient temperature), maximum temperature (operating) and the test temperature for this section – please refer to the standard.

	X	Y	
Temperature rise (measured) °K	35	35	
Maximum operating (ambient) °C	25	40	
Maximum temperature (operating) °C		60	75
Section 4.2.6 test temperature °C		70	85

These criteria could be applied should a plastic part break.

EN 60950 Standard reference	Synopsis
4.2.8	Mechanical strength of cathode–ray tubes. The standard requires cathode–ray tubes to comply with IEC 65.
4.3	CONSTRUCTION DETAILS
4.3.1	Equipment which can be adjusted to suit different primary voltage supplies must require a tool to change the settings if the incorrect setting could create a hazard.
4.3.2	Applies to accessible control devices whose adjustment could create a hazard and requires that those controls be accessible using a tool.
4.3.3	Reserved for future use.
4.3.4	Relates to equipment producing dust using powders, liquids or gases.
4.3.5	Requires knobs, handles, grips, levers to be reliably fixed, if in working loose they could result in a hazard. It prohibits the use of sealing and adhesives. The test requirements for compliance are contained within this section of the standard.
4.3.6	Relates only to driving belts and other mechanical couplings that provide electrical insulation.
4.3.7	Requires that sleeving used to provide supplementary insulation must be positively retained.
4.3.8	Deleted.

Guidance notes

Ensure that CRTs are treated as safety critical components.

Most commonly, we would expect equipment rated and set for 110 V to fail catastrophically when connected to 230 V. If the only effect of this is to blow the mains fuse then, within the meaning of this standard, we do not create a hazard and that situation is acceptable.

If we connect equipment configured for 230 V operation to a mains supply of 110 V and find that under fault conditions there is insufficient current to operate the protection devices, we may find that this scenario is a greater hazard within the meaning of this standard.

It is usually simpler to provide a switch which requires a tool to change the settings.

Assessment for hazards would take part under abnormal testing. See section 5.4 *et seq.*

This section contains vital mechanical and electrical design criteria.

In determining whether working loose could create a hazard, consider the effects of rotation and how that may affect interlocks or indicating the status of the equipment.

It must not be possible to replace the belt, the coupling or any other parts in such a way as to introduce a hazard.

The design must be such that incorrect assembly or replacement of parts cannot introduce a hazard (e.g. replacing plastic screws with metal screws).

Tie wraps can be used to restrict the movement or to clamp the sleeving on to other parts.

Similar precautions must be taken where sleeving provides basic insulation.

When fitting sleeving, avoid cuts along its access which provide a break in insulation (e.g. do not slit it along its length to aid fixing).

Small gaps in reinforced insulation are no longer allowed.

Deleting this section has created a new requirement and brought this

EN 60950 Standard reference	Synopsis
4.3.9	Requires that equipment should permit any nut, bolt, screw, washer, spring or similar part to fall out of position and not compromise creepage distances and clearance distances over supplementary or reinforced insulation. You are allowed to assume that two independent fixings will not become loose at the same time. Parts fixed by locking nuts or nuts with self-locking washers or other means of locking will not become loose unless these screws or nuts must be removed to replace a power cord. Wire breakages must be considered (unless insulation is clamped also) whether these are soldered, crimped, clamped, or held in a terminal. You are allowed to assume that short wires will not come away from a terminal if, after loosening the screws of the terminal, the wire remains in place.
4.3.10	Deleted
4.3.11	Requires insulation material that is exposed to oil, grease or other hydrocarbons to be resistant and not suffer degradation of its properties.
4.3.12	Considers other hazards such as ionizing radiation, ultraviolet (UV) light, laser, flammable liquids and gases.
4.3.13	Requires electrical or mechanical screwed connections to withstand normal mechanical stresses if their loosening or failure could affect safety.
4.3.14	Begins the section on fire enclosures.

Guidance notes

harmonized standard into line with the section of North American standard UL 1950 §4.3.8.

Tie wraps can be used to restrict the movement or to clamp the sleeving on to the source of hazard.

Areas to consider are single fixings and locking fixings that must be removed to replace power cord. The wire mesh described in section 4.4.6 (no greater than 2 mm by 2 mm with a wire of not less than 0.45 mm) could be of great value here.

The fracture of all strands in a multistrand wire is considered a single failure, thereafter the wire may be bent without forcing to its extreme positions and contact between hazardous live parts and SELV and other hazardous combinations would result in a failure. Positive clamping and retention using cable ties is acceptable to limit movement of wires.

If the wire is short and rigid then two terminal screws would need to be loosened before it could be removed.

Deleting this section is a relaxation of previous requirements.

It is best to avoid these materials. Where they are used in current designs we should plan to replace them with more suitable materials.

If these sources are present within the equipment we must ensure that insulation and other materials are not degraded by them.

The general solution to this requirement is to use locking nuts, spring or star washers, nylok or special nuts.

Within this standard, the phrase 'fire enclosures' has a very special meaning: it may apply to the entire enclosure, the base of the enclosure, the side of an enclosure or locally to any small part of any surface of the enclosure.

Within the meaning of this standard, a metal plate covering part of the base of an equipment may fulfil all the requirements of a fire enclosure.

The need for a fire enclosure is determined by components and the local environment within the equipment.

Before designing the internal layout of equipment, identify all components and assemblies which will need a fire enclosure. This will significantly reduce design, hardware and compliance costs.

Before proceeding through this section read the flammability guidance notes in the easy guide.

EN 60950 Standard reference	Synopsis
4.3.15	This section covers some of the requirements for the tops of fire and electrical enclosures. Openings directly over bare parts at hazardous voltage may not exceed 5 mm in any direction. May not exceed 1 mm in width regardless of length. Alternatively, opening may be designed similar to the sketches in Figure 8 of the standard to prevent direct vertical entry of a falling object.
4.3.16	This section covers similar requirements for side openings in fire and electrical enclosures An object falling within the profile of Figure 10 of the standard must not fall onto bare parts at hazardous voltages.
4.3.17	Requires that operator and service accessible plugs, sockets and connections should not create a hazard if they are mis-mated.
4.3.18	Relates to equipment intended to plug directly into a wall socket outlet.

Guidance notes

Flammability requirements are covered between this section and section 4.4.8.

Therefore be extremely cautious when designing enclosures and fire enclosures, and ensure they comply with *all* of these requirements.

Above all – *take your time.*

The requirements of the following sections are not easy reading and the flow of the requirements may not best suit the needs of the designer.

Please refer to the following sections in Chapter 8, they will provide a step-by-step approach to determining where we need a fire enclosure and how to go about its design:

- 'When a fire enclosure is necessary'
- 'Construction for the bottom of a fire enclosure'
- 'Construction for the side of a fire enclosure'
- 'Construction for the top of a fire enclosure'
- 'Flammability requirements for components, parts and assemblies'

Please read the standard in conjunction with these documents.

Imagine that the entire top of the enclosure was missing and that you are looking vertically down into the equipment. Can you see any bare parts that are at hazardous voltage, or that present an energy hazard?

If there are no bare parts at hazardous voltage the following sections do not apply, proceed to the next section.

A 4.9 mm diameter hole is acceptable. If we use a 5.0 mm hole then 50 per cent of production will be non-compliant.

A slot of 0.9mm running the entire length or height or around the diameter of the equipment is acceptable: however, consider the mechanical strength tests (250 N and ball impact) and the effect that these tests may have on the dimensions of the slot.

Do not exceed the dimensions given in the standard.

The apertures should not exceed 5 mm in any one dimension nor exceed 1 mm in width regardless of its length (see section 3.4.15 above).

To prevent objects falling into the enclosure louvres may be added as shown in Figure 9 of the standard.

Please refer to Chapter 8.

It is allowable to restrict cables and looms so that they may only be connected to the appropriate connector. Alternatively, other methods are suggested in the standard. Note that warning markings are particularly applicable to service personnel.

This applies to small SELV power supplies used for such equipment as calculators, telephones etc.

Detailed requirements are contained within the standard.

EN 60950 Standard reference	Synopsis
4.3.19	Build-up of liquid and pressure.
4.3.20	Considers earth faults in heating elements used in Class I equipment.
4.3.21	Requires equipment using lithium and similar cells to be designed to prevent reverse polarity insulation of the battery cell and to prevent forced charging or discharging to create a hazard.
4.4	RESISTANCE TO FIRE
4.4.1	Methods of achieving resistance to fire. There are two methods: the first is to select components and materials to minimize both the possibility of ignition and the spread of flame; the second is to carry out full fault simulation on all components within the system. This option is included within section 5.4.6.
4.4.2	Minimizing the risk of ignition. This contains references to other parts within this section. In particular, note the requirement that 'where it is not practical to protect components against overheating under fault conditions, the components shall be mounted on materials of flammability Class V1 or better and shall be separated from less fire resistant material by at least 13mm of air.
4.4.3	Flammability of materials and components.
4.4.3.1	General.
4.4.3.2	Flammability. The general requirements for materials and components are: • they shall have a flammability class of V-2 or better; • they shall have a flammability class of HF-2 or better; • they shall pass the flammability tests of clause A2.
4.4.3.4	Exemptions. The exemptions to the above section (4.4.3.2) are lengthy and detailed – please refer to the standard.

Guidance notes

Requires protection from build-up of excessive pressure where liquids are employed.

If an earth fault could energize the heater and cause overheating, then this failure must be investigated. Detailed requirements are contained in this section of the standard; see also the requirements for Abnormal Testing.

The potential hazards are explosions and fire. In particular faults which could cause the batteries to be charged must be evaluated.

Please refer to Chapter 8.

Fault simulation can be expensive in terms of time and money. The designer is far better to concentrate his efforts on designing (or even over-designing) the equipment to comply with this section of the requirement.

Hence we can shield components which overheat or ensure an air gap of at least 13 mm. Make sure that components cannot be bent or displaced (e.g. wirewound resistors mounted on a printed circuit board can be bent towards flammable material: 13 mm clearance is a minimum).

Of key importance within this section is the requirement to consider the cumulative effect of small parts which may propagate fire from one part to another.

Before reading this, check the exemptions in the next section.
    It will be more cost effective to select an appropriate plastic part from a recognized source than to enter a perpetual regime of conformity testing.

Note that components meeting the flammability requirements of the relevant IEC components standard are exempt from other flammability requirements.

EN 60950 Standard reference	Synopsis
4.4.3.4	Wiring harnesses. This section requires materials to have a flammability class of V-1 or better or to comply with the relevant IEC Standards. Alternatively harnesses constructed with PVC and other materials are exempt.
4.4.3.5	Cord anchorage bushings applied over PVC jacketed power cables shall have flammability class HB or better.
4.4.3.6	Air filter assemblies. The general requirement is that air filter assemblies be constructed of materials of flammability class V-2 or better but the following relaxations are possible: • air filter assemblies in air conditioning systems that are not intended to be vented outside of fire enclosure need not comply with this requirement; • air filter frames constructed of flammability HB and separated from electrical parts which could reach high temperatures under fault conditions must be separated by a 13 mm air gap or a solid barrier of flammability class V-1 or better; • air filter assemblies located inside or outside a fire enclosure provided that the filter materials are separated by a metal screen from parts that could cause ignition; • air filter assemblies located externally to the fire enclosure constructed of materials of flammability HB or better.
4.4.4	Materials for enclosures and for decorative parts. This section contains the general requirements. In particular: metal; ceramic parts; and heat resistant, tempered, wired, or laminated glass will comply with the requirements without test. Mechanical enclosures, electrical enclosures and parts of enclosures located externally to fire enclosures, and decorative parts shall be of flammability class HB or better. Small external decorative parts that would contribute negligible fuel to the fire are exempt from this requirement. The fire enclosure required for equipment that is movable and having a total mass not exceeding 18 kg is

Guidance notes

Note also that small electronic parts are exempt provided that they are mounted on material (e.g. a pcb of flammability class V-1 or better).

Equipment wire marked UL or CSA or VW-1 will comply with this requirement.

This is the same flammability classification for decorative parts.

Note that V-2 material must not form part of a fire enclosure.
Please see exact wording in the standard.
The metal screen must comply with the requirements in section 4.4.6 of this standard, or bottoms of fire enclosures (e.g. mesh not greater than 2 mm by 2 mm and of wire diameter not less than 0.45 mm or a plate with holes meeting the size and spacings of Table 15).

Remember that the outside of enclosures must survive the ball impact test to qualify as a fire enclosure.
It follows that if the equipment does not require a fire enclosure then the mechanical enclosure may be of flammability class HB.
This includes name plates, key caps, knobs etc.
Movable means not fixed. For fixed equipment the mass is 18 kg maximum – for equipment approaching this mass consider using 5 V material.
The fire enclosure of fixed equipment needs the same flammability rating as equipment weighing more than 18kg – UL94–5 V.
This includes such things as commutators, unenclosed switch contacts etc.
When considering placing items within these 13 mm limits, consider the

EN 60950 Standard reference	Synopsis
	required to be V-1 or better. Alternatively it passes the test in clause A1 of the standard. For movable equipment having a total mass exceeding 18 kg and all fixed equipment, fire enclosures should have a flammability rating of 5 V. Alternatively they may pass the test of clause A1. Enclosures and parts of enclosures located within 13 mm of arcing parts are required to pass the test of clause A3. Enclosures or parts of enclosures that are located within 13 mm of parts which could attain excessive temperatures shall pass the test of clause A4. Components which fill an aperture are not evaluated for compliance with the flammability requirements of the fire enclosure provided that they comply with the relevant IEC component standard.
4.4.5	Conditions for fire enclosures.
4.4.5.1	Defines the list of components requiring a fire enclosure. Note the requirement for wound components such as transformers, solenoids and relays.
4.4.5.2	Defines the components that do not require a fire enclosure.
4.4.6	This includes a major exemption that equipment which can only be energized if an operator is in attendance is exempt for all needs for a fire enclosure. This section refers to Figure 11 of the standard. The 5° angles shown in Figure 11 are significant because these have the lines marked with the maximum extent that flaming droplets or particles are considered to be emitted. The section continues to define the requirements for the bottom of a fire enclosure.
4.4.7	Doors and covers in fire enclosures If part of a fire enclosure consists of a door or cover leading to an operator access area one of the following shall apply: The door or cover shall be interlocked to comply with section 2.8 of the standard, this requires interlocking of all hazardous voltages and energy levels. A door or cover intended to be routinely opened by the operator shall comply with both of the following conditions.

cost of compliance versus the design impact of providing a 13mm (minimum) air gap.

The standard allows us to use IEC compliant parts as an integral part of a flame enclosure. Remember that all impact tests must still be conducted.

Please refer to Chapter 8, 'When a fire enclosure is necessary'.

Please refer to Chapter 8, 'When a fire enclosure is necessary'.

Please refer to Chapter 8, 'Construction for the bottom of a fire enclosure'.

Note particularly the use of baffles: baffles are frequently very large and may extend over half the depth of an enclosure to allow access of cables from beneath the equipment.

It may be easier to view this from the opposite perspective of the operator's point of view. If the operator has access to a cover or a door in a fire enclosure then the special conditions apply.

This is usually achieved by using hinges.

Instructions must be in local language. It is advisable that fixings are captive. Otherwise ensure that they cannot be dropped into hazardous live areas.

Remember that earth bonds may be used only for the purpose of earth bonding and should not be used to retain covers in this operator application.

EN 60950 Standard reference	*Synopsis*
	It shall not be removable from the fire enclosure by the operator. It shall be provided with a means to keep it closed during normal operation. The third alternative applies to a door or cover intended only for occasional use by the operator which may be removable, provided that the equipment instructions include directions for correct removal and replacement of the door or cover.
4.4.8	Flammable liquids. This section has clear and explicit requirements for the use of flammable liquids within equipment.
5.0	THERMAL AND ELECTRICAL REQUIREMENTS
5.1	HEATING. In normal use, the equipment and its component parts shall not attain excessive temperatures. When determining temperatures under normal load conditions, we must consider whether the equipment is continuously operated, intermittently operated or operated for a specified short term duration. To these factors we must add the worst case of voltage and frequency. The requirements for maximum temperatures of items include: that thermal cut-outs do not operate; and ceiling compounds do not leak out during normal operation.

Guidance notes

Note the maximum capacities allowable in the first paragraph, and the dependence upon flashpoint of the material.

The volume of liquid allowable will have a major impact upon the design of the equipment. Similarly it may be valuable to add explicit instructions to equipment designed for use with liquids with flashpoints above 149°C to warn operators and service personnel of the limitations when using lower flashpoint liquids.

This is checked by measurement. Temperature rises of components above ambient air temperature and the maximum operating temperature of the component, wire, subassembly etc. must be considered when specifying the maximum ambient temperature in which the equipment can operate (e.g. if the laboratory temperature is 20°C and the component's temperature is measured at 65°C and its maximum operating temperature is 85°C, then the component temperature rise is 45°. This we subtract from the maximum component temperature of 85° and conclude that the maximum ambient temperature in which the equipment can operate is 40°C.

Temperature rise = measured temperature minus ambient temperature = 65 − 20 = 45°C.

Maximum operating ambient = maximum specified component temperature minus temperature rise which equals 85 − 45 = 40°C.

With linear power supplies, worst case dissipation may be at 50 Hz and maximum operating voltage. This is the worst case for transformer magnetic and resistive losses and also maximum dissipation on regulating components.

With switched mode power supplies, it is less easy to identify worst case operating conditions and these must be found by experiment (a power meter is extremely useful in determining worst case conditions).

When testing large or complex items of equipment, it is permissible to test components in isolation under the conditions that they would operate within the equipment.

It is vital that installation instructions contain explicit instructions or equipment installation. This is the only acceptable method of ensuring equipment is installed correctly (e.g. clearance and ventilation

EN 60950 Standard reference	Synopsis
	Table 16 temperature rise limits Part 1. This table defines the maximum temperature rise limits for a variety of components. This includes insulation, terminals for stationary equipment etc., parts in contact with flammable liquid, and components. Table 16 defines the maximum temperature for operator contactable parts.
5.2	EARTH LEAKAGE CURRENT
5.2.1	General. This section differentiates the test requirements for IT power systems and the test required for TT or TN power systems. The maximum leakage currents permissible are listed in Table 17 and the basic test requirements for Class A and Class II equipment are included in the text.
5.2.3	Covers the tests for single phase equipment.
5.2.4	Describes earth leakage measurements for three phase equipment. It includes single fault failures in EMC (EMI) protection.
5.2.5	Covers the requirements for Class I equipment whose earth leakage current exceeds 3.5 mA. The requirement is that the equipment shall be stationary and permanently

requirements, maximum ambient temperature, limitations on placing in direct sunlight, routine cleaning of air filters, overload protection fitted in building wiring).

First remember that these are temperature rises against an assumed temperature of 25°C (add 25°C to all of these temperature rises for the maximum allowable temperature during normal operation at maximum ambient temperature).

Wire may be marked with a temperature rating, this is referred to as 'T-marking', e.g. 'T-25'. Note also that wire PVC, rubber etc. which has no temperature marking has an assumed maximum temperature rise of 50° (50 + 25 = 75°C maximum temperature).

These are classified into areas which are unlikely to be touched, will never be touched, and may be touched regularly. Note in particular that external surfaces which have no single dimension exceeding 50 mm and which are not likely to be touched in normal use may have a temperature rise of up to 75°.

This means the actual temperature may be 100°. A warning notice should be included if this is the case: take particular care that the label and adhesive are suitable for this high temperature application.

When considering areas of the equipment which exceed temperature limits the two considerations are that unintentional contact is unlikely and that there is adequate warning that the part is hot. (Remember language requirements or use of international symbols.)

This important requirement is often overlooked in product design.

When making earth leakage current measurements take extreme care as incorrect connection, or a fault in the equipment, can cause substantial currents to flow.

The maximum earth leakage current is generally at the highest voltage. Equipment contains switched–mode devices which may exhibit maximum leakage currents at some intermediate voltage.

Tests must be made with switches in on and off positions. If equipment is for use where supplies are not polarized, measurements should be made with live and neutral transposed (this will simulate a non-polarized plug being inserted 'upside down').

The standard requires the worst case failure to be considered (e.g. that any single point connection that can cause an imbalance across the phases must be evaluated).

The following is not a requirement of this standard but is recommended by the author. It is very helpful to include typical earth leakage current data within the servicing or installation instructions. This will be of great value

EN 60950 *Standard* *reference*	*Synopsis*
	connected to building wiring, or that it is connected via an industrial type plug (connector type B). It is not acceptable under any circumstances for earth leakage current to exceed 5 per cent of the total input current (per phase). All equipment with earth leakage exceeding 3.5 mA must carry a warning label (wording is suggested in this section). Remember language requirements and include in this warning in-servicing and operator instructions.
5.3	ELECTRIC STRENGTH
5.3.1	Deals with the general requirements that insulating material shall be adequate. Identifies the test requirements as Annex C, section 3 for safety isolating transformers and for other equipment in accordance with section 5.3.2.
5.3.2	This section deals with test methods and test voltages which are defined in Table 18, parts 1 and 2.

Guidance notes

to your customers as it will highlight areas of care which may require special attention (e.g. within rack mounted equipment make sure that earth conductors are a minimum of 1 mm²) in office environments and ensure that equipment connected to multiway sockets do not violate local wiring regulations and health and safety requirements.

This will lead you to reading Table C2 'electric strength tests' which define the requirements for Operational, Basic, Supplementary and Reinforced insulation. The figure in this document is not an exhaustive treatment and contains fewer examples than the standard but may provide a clearer understanding to the needs for insulation.

IMPORTANT
If in doubt use Reinforced insulation. This may prove slightly more costly but will reduce compliance, proving costs and the likelihood of an even more costly non-compliance.

Read the notes for Table C2 carefully. They are written concisely and will need re-reading several times – it may help to read these in the reverse order starting with note 8. This will cover the exclusions and definitions first.

Ensure that test methods and procedures deal with combinations of Reinforced and Basic insulation. It is essential to isolate or remove components and circuit elements protected by basic insulation before conducting tests for Reinforced insulation.

It is helpful to identify printed circuit boards and circuit layouts with coloured markers representing different insulation types – this can be of great value during test and will help avoid damaging the unit under test.

Remember that equpment used on a nominal 230V mains supply will have a working voltage of at least 254 V therefore use the 250–1000 V r.m.s. column of Table 18, Part 1. Note also, for the row marked 'Basic supplementary' that the requirements apply to Basic insulation and Supplementary insulation as to individual insulation elements and tests.

Basic insulation is tested in accordance with Table 18, Part 1, the tests for Supplementary insulation are identical to that of Basic insulation. Basic and Supplementary are not to be tested as though they were a combined insulation to the requirements of Table 18.

Full test requirements are included in section 5.3.8.

EN 60950 Standard reference	Synopsis
5.4	ABNORMAL OPERATING AND FAULT CONDITIONS
5.4.1	Equipment shall be so designed that the risk of fire and electric shock due to mechanical or electrical overload or failure, or due to abnormal operation or careless use, is limited as far as practicable. After abnormal operation or a fault the equipment shall remain safe for an operator within the meaning of this standard but is not required to be in full working order. It is permitted to use fusible links, thermal cut-outs, overcurrent protection devices and the like to provide adequate protection. The standard allows us to carry out certain testing at component or subassembly level.
5.4.2	Under overload, locked rotor and other abnormal conditions, motors shall not cause hazards because of excessive temperature. Methods for achieving safe operating conditions could be the use of inherent or externally protected motors. Using a motor powered by a secondary circuit. Current sensing motor current. The use of an integral thermal cut-out. The final suggestion within the standard is to detect that the mechanism is not performing correctly and to disconnect power from the motor (e.g. the cut-outs are tripped if a mechanism fails to operate within a certain time, or operating period).
5.4.3	Transformers shall be protected against overload, for example by: ● overcurrent protection; ● internal thermal cut-outs; ● use of current limiting transformers. The compliance requirements for transformer overload are detailed in clause C1 of Annex C.

Guidance notes

The phrase 'as far as practicable' is important and may be the key to legal defences of designers and test engineers. Unfortunately the standard does not contain an absolute definition of what would be considered practicable.

No doubt this will be left to courts in the future.

Here we are dealing with single fault conditions and foreseeable misuse. We must also consider any subsequent faults following an abnormal fault condition.

Remember to consider the need for fuses in neutral lines, adequate breaking capacity for overcurrent protection devices and suitable safety approvals for all safety critical components. Remember also that self-resetting protective devices (e.g. thermal cut-out) must not create a hazard under servicing and normal operating conditions.

Considerations must be fire, physical injury to operator and service personnel, and electric shock hazard (following breakdown of insulation).

Check that the motor is specified for correct operating voltage and frequency range.

Here basic insulation may be put in place by a suitable transformer.

This is particularly useful for large and efficient motors where overload causes a substantial increase in motor current.

This may be useful for marginal overload conditions but check that thermal time constants do not make this sensor unsuitable for locked rotor conditions.

The test requirements are detailed in an Annex B of the standard.

Consider overloads caused by single component failures within the secondary circuits, failures of basic or supplementary insulation to either or both the live and neutral of the transformer, and overloads applied externally to the transformer and its associated circuitry (e.g. overloading a power supply).

Check that fuses or similar protection devices will operate and protect the transformer under all voltage and supply frequencies. (NB. In small transformers the difference between the primary current under normal conditions and under short-circuit secondary conditions may be inadequate to blow fuses reliably.)

As with motors (see section 5.4.2) these are useful in providing overload protection. Their use for overcurrent protection may be limited by thermal time constants within the equipment.

This is a useful solution to problems associated with small transformers:

EN 60950 Standard reference	Synopsis
5.4.4	Defines the test to be applied to operational insulation. This requires creepage and clearance distances to comply with the requirements of section 2.9. Alternatively they may be tested to the appropriate electric strength in 5.3.2. Alternatively the operational insulation may be short-circuited to demonstrate that such a failure would not create a hazard (the criteria are defined in the standard).
5.4.5	Deals with electromechanical hazards and failures in secondary circuits. The standard contains detailed test and acceptance criteria.
5.4.6	Details the fault conditions that must be simulated in components and circuits that have not yet been covered by the standard and includes: • faults in any component in primary circuits; • faults in any component where failure could adversely affect supplementary or reinforced insulation; • where the flammability requirements contained within sections 4.4.2 and 4.4.3 are not met faults must be simulated on all components within the equipment; • faults arising from connection of the most unfavourable load impedance to terminals and connectors that deliver power or signal outputs from the equipment, other than mains power outlets; • where multiple outlets share the same internal circuitry the test is made only for one sample outlet; • where primary components comply with section 5.4.4 option A (have adequate creepage and clearance distances) there is no fault simulation. As with other sections please read the requirements within the standard.

current limiting transformers with safety approvals in place are available from many stockists.

All other things being equal it is better to provide adequate insulation in all conditions. However, the standard recognizes that this is not always practicable and allows the designer either to prove, by appropriate electric strength tests, that operational insulation is adequate for the purpose; or to demonstrate that there are no consequential flammability or electric shock hazards following a failure of an operational insulation.

The duration for tests is normally until steady-state conditions are achieved. This is because most equipment may be energized without an operator being present.

Normally short-circuit and open-circuit are the only conditions considered. However, if a component has another mode of failure which may create a hazard that must be investigated.

Usually this means overheating supplementary or reinforced insulation and should be considered at component level and within wiring looms (e.g. where a wirewound resistor may overheat check that adjacent wiring is positively fixed away from the source of heat, and that the resistor is positively fixed and cannot be bent or displaced so that it may damage insulation)

Designing to meet the flammability requirements of this standard is fundamental to containing compliance costs.

Inadequate knowledge of this safety standard can cause massive overspend in development and delay time to market by many months.

Simulating faults on every component is, even for simple equipment, a major exercise.

Remember that the equipment must be operational before commencing the next investigation.

It is a useful idea to measure the maximum output voltages and currents that can be drawn from any signal outlet early in the design phase. Provide adequate protection not only improves safety compliance but protects the equipment from foreseeable abuse.

This removes the need to test outlets which are paralleled together (e.g. a databus).

Provided that these components have appropriate safety approvals to demonstrate conformity to the relevant IEC standard (see section 1.5.1 'Components').

EN 60950 Standard reference	Synopsis
5.4.7	Contains the key requirement that equipment shall be tested by applying any condition that may be expected in normal use and foreseeable misuse.
5.4.8	Contains the detailed requirements for unattended equipment which has thermostats, thermal limiters, or thermal cut-outs, or that has a capacitor not protected by an overcurrent device.
5.4.9	Contains the acceptance criteria including the requirement that if a fire occurs it shall not propagate beyond the equipment.
5.4.10	Details special testing or thermoplastic parts on which hazardous voltage is mounted directly.
6.0	CONNECTION TO TELECOMMUNICATION NETWORKS
6.1	REQUIREMENTS. This applies to equipment that is directly connected to telecommunications networks (TNV circuits) to provide adequate protection to ensure compliance with the requirements for TNV circuits and protection against electric shock for service personnel and other users of the telecommunications networks from hazards in the equipment.
6.2	TNV CIRCUITS AND PROTECTION AGAINST ELECTRIC SHOCK
6.2.1	TNV circuit characteristics and requirements. This details what the designer should anticipate under normal and single fault conditions on a TNV signal.
6.2.1.2	This section applies to TNV circuits normally operating in excess of the SELV limits. There shall be basic insulation or better between TNV circuits and unearthed operator accessible conductive parts. TNV circuits and unearthed SELV parts.

Guidance notes

This must include the effects of voltage variation.

Please refer to the standard for detailed requirements and test criteria.

Note the requirement: '... equipment ... with more than one ... each is short-circuited, one at a time'.

If thermal cut-outs are needed in a safety critical application then fit two 'approved' cut-outs and connect them in series. This should reduce compliance testing significantly.

Remember impact on fire enclosures etc.

The equipment shall not emit molten metal. Protective glasses and other equipment are necessary when carrying out abnormal testing.

Please refer to this section for the remainder of these requirements.

If using this method of construction test samples very early in design: this will reduce the likelihood of an expensive redesign late in the design phase.

That hazards are not introduced into the TNV circuits and the telecommunication network.

TNV circuits may have a maximum d.c. voltage of over 70 V and a maximum peak voltage of 120 V. There is an obvious difference between the requirements for SELV and the allowable voltages for TNV circuits. These are resolved in the following sections.

These are the voltages against which the designer must provide adequate insulation to protect his equipment, the service engineer and the operator. Note that any connection to a TNV circuit will not exceed 1500 V peak – see Figure 15 within the standard.

Note Figure 15 of the standard.

EN 60950 Standard reference	Synopsis
6.2.1.3	Applies only to TNV circuits normally operating in excess of the SELV limits. This section modifies the SELV limits under single fault conditions.
6.2.1.4	Requires TNV circuits to be separated from hazardous voltages by either double or reinforced insulation by basic insulation together with a protective screen or by designing to comply with section 6.2.1.5.
6.2.1.5	Allows a TNV circuit to be supplied by a d.c. voltage not exceeding 120 V produced by rectifying an a.c. voltage of not greater than 50 V provided that it is separated from hazardous voltage by double or reinforced insulation or by basic insulation with a protective earthed screen and that the voltage limits in Figure 15 of the standard are not exceeded in the event of a single insulation failure or a component failure.
6.2.2	Protection against contact with TNV circuits. This requires adequate protection against contact with bare conductors and TNV circuit parts carrying voltages exceeding the SELV limits under normal operating conditions.
6.3	PROTECTION OF TELECOMMUNICATION NETWORK SERVICE PERSONNEL AND OTHER USERS OF THE TELECOMMUNICATION NETWORK, FROM HAZARDS IN THE EQUIPMENT.
6.3.1	Protection from hazardous voltages. Circuitry intended to be connected to a telecommunication network shall comply with the requirements for TNV circuits. This applies whether or not such circuitry is operator accessible prior to the connection to the telecommunication network.
6.3.2	Use of protective earthing. Protective earthing of Class I equipment shall not rely on the telecommunication network.

This is a short-term transient relaxation only. The steady-state condition still applies.

Note the voltage limits within Table 15 within the standard. Layout and single fault conditions must be considered.

Please read this section of the standard very carefully.
Detailed analysis and testing are required to ensure compliance.

The designer may wish to confirm the voltage limits in Figure 15 when designing the safeguards.
Note, however, that the following are exempt from this requirement:

- contacts with connectors that cannot be touched by the test probe detailed in Figure 16 of the standard;
- equipment intended for installation in a restricted access location.

Consider that any equipment failure which can introduce a hazardous voltage onto the network may affect all equipment connected to that network
Visualize a simple office network of PCs, printers, modem etc. Now imagine a fault that raises the network to a hazardous voltage.
All exposed data connections and keyboards could become a potential source of hazard.
It is to prevent this and other scenarios that EN 60950 was developed.

Hence Class I equipment must be provided with its own earth.

EN 60950 Standard reference	*Synopsis*
6.3.3	Does not apply to equipment intended to be installed by service personnel.
6.4	PROTECTION OF THE EQUIPMENT USER FROM VOLTAGES ON THE TELECOMMUNICATION NETWORK
6.4.1	Separation from telecommunication network conductors requires equipment to provide adequate electrical separation between the port provided for telecommunication network connection and the following: • unearthed conductive parts and non-conductive parts expected to be touched or held during normal use; • parts and circuitry that can be touched by the test finger in Figure 19 of the standard but excluding contacts that cannot be touched with the test probe defined in Figure 16 of the standard circuitry which is provided for connection to other equipment
6.4.2	Details the compliance testing and permits two test methods contained in sections 6.4.2.1 and 6.4.2.2.
6.4.2.1	Defines the impulse tests required to be injected.
6.4.2.2	Details electric strength tests and permits a.c. or d.c. testing.
6.4.2.3	Compliance criteria. Insulation breakdown shall not occur.

Guidance notes

This means installed by an operator.

If the equipment is pluggable type A or has a warning (in the local language) stating that protective earthing is required for safety reasons, there must be supplementary insulation between the telecommunication network and the protective earthing. Please check the precise wording and detailed testing required for capacitors, surge arresters etc.

This is to provide adequate isolation to SELV circuits and limited current circuits from the telecommunication network.

This requires isolation barriers to be adjacent to the TNV circuits. It does not, however, apply to circuitry which carries TNV signals.

Note insulation requirements between the TNV signals and adjacent circuitry throughout the loom and in particular creepage and clearance distances at connections.

Please read these and the following sections carefully.

The standard is very clear in this requirement.

The standard is very clear in this requirement.

Breakdown is defined here and checks for insulation degradation are defined.

6   Achieving compliance

The CE Marking Directives require a well-defined level of documentation to be available for inspection. In the case of present safety legislation, this documentation does not need to exist as a report, provided that it can be generated and made available should the enforcement authorities request it.

To comply with the CE Marking Directives is not difficult and involves several simple steps: but there are a lot of detailed requirements and it is easy to make mistakes or to waste time and money.

The following steps outline a path to enable companies to carry out compliance work efficiently and correctly; and to build the legally required 'Data File' and to ensure that all subsequent products manufactured are compliant.

Understanding

We must understand the business, the people and the product. Without this level of knowledge it is not possible to determine how best to proceed. With our new understanding we set about designing products to an appropriate safety standard.

Product review

We complete a 'competent body' style report – explaining the interpretations is only part of the task. We learn a new level of awareness is necessary to view the product (as the Directive requires) from the perspective of the phrase 'reasonable use and foreseeable abuse'.

Product test

We design a test plan that will reflect likely operation under normal *and* fault conditions, then we detail the test procedures for safety proving ('typetesting') and production testing.

Manufacturing control

The work so far has proven that the product reviewed and tested complies with the Low Voltage Directive – we must now put controls in place to ensure that all subsequent products will also comply. Although a quality system may assist in this task, a quality system without appropriate manufacturing controls will fail to ensure product conformity and will put at risk all of the work and training carried out.

Ongoing support – the Internet etc.

There is always an occasional feeling of vulnerability and being 'alone'. Your local EMC club will be a good starting point. Many of these clubs are beginning to include the Low Voltage and other Directives within their scope.

Internet provides web pages, newsgroups and bulletin boards that offer free advice: the best newsgroup I have located can be found at 'sci.engr.electrical.compliance'. The FAQs (frequently asked questions) are published frequently and are well worth the several minutes' download time. Similarly there is a free downloadable safety help file on http://www.gkcl.com. This will run on any IBM compatible under Windows V3.1 or above. And remember, there is an old saying that 'free advice is worth every penny'.

The fundamental strength of the Internet is that everyone can ask questions: but this strength is also the ultimate weakness of free access pages – anyone can read your personal confession, anyone can comment and not all contributors have a valid opinion. The best advice is to find a test house or consultant that you trust and then build an ongoing relationship with them. They help you to develop your business and engineering practices to meet not only the CE Marking New Approach Directives but other national and international standards as well.

Speaking from my personal experience I have found the long-term relationships with client companies most rewarding and take great pride and personal satisfaction in the (now annual) visits to the longest clients.

Compliance proving

Once we have designed our compliant equipment and demonstrated that it is compliant with the essential requirements of the LVD we can now begin to sell and place the equipment upon the European market. Compliance proving requires a combination of physical review of the equipment, testing and documentation. The physical review takes a detailed knowledge and understanding of the standard, and of the equipment including its likely use and its foreseeable misuse.

Detailed testing: normal conditions and abnormal conditions

These topics are beyond the scope of this book and most short courses.

Documentation and the Data File

It is sensible, therefore, to construct a Data File as you proceed. The following is the list of documentation usually required:

- A general description of the product.
- Conceptual design sketches and calculations.
- Manufacturing drawings.
- Schemes for components.
- Schemes for subassemblies.
- Schemes for circuits.
- Descriptions and explanations necessary to understand the above drawing, schemes and the operation of the product.
- A list of the European harmonized standards which have been applied in full or in part.
- Descriptions of the solutions adopted to meet the essential requirements of the Directive(s) where the harmonized standards have not been applied.
- Results of design calculations.★
- Results of examinations and tests carried out.★
- Test reports.★
- Approval certificates and reports issued by independent agencies.★

(★ These items are fundamental in proving compliance and will comprise the bulk of the technical file.)

This level of information allows the product to be assessed to see if it conforms to the minimum required level of safety. Many companies consider it worthwhile to exceed these minimum levels for a variety of reasons: avoiding product liability claims, marketing to countries with different requirements, company image, reducing the risk of delay in a product launch.

During design – components

You should document and maintain safety data throughout the entire project. Designers should gather safety information (copies of certificates from independent agencies or test houses) for all safety critical components used. This will save the exercise being repeated later and may avoid discovering that a critical component has no safety approvals and that an alternative must be found at a late stage. It is far easier to get this information from the sales representative before ordering the parts. Never trust a verbal assurance – always get you hands on a copy of the safety certificate and check the date and other data for validity. It will create a great deal of trouble if a component is found to be unapproved later in the development cycle.

During design – testing

Safety testing should take place throughout the design phase. Consider the alternative: who would commit many millions of pounds to designing an aircraft if they could not be sure that it would fly? Equally, who would commit their entire research budget to design a product and not be sure that it can legally be sold until it has been tested?

Companies should make more effort to safety test during the design phase to give a greater confidence that their product will pass formal testing at the end of the development phase.

Testing

Here there are two general ways of proceeding:

1. Use an external test house.
2. Develop the resources in-house.

Option 1 is the easiest and one simply pays the bill.

Its disadvantages are that the designers do not gain the best benefit, that the knowledge base does not grow within the company, and there may be no method for maintaining compliant products. Therefore in the long term costs are ongoing.

Option 2 is the more difficult to achieve – training is paramount – 'little and often' is the key to successful product safety training. But the short-term costs are significantly higher than option 1.

However because we are training our designers, we will find fewer non-compliances, there will be less redesign and the design costs will decrease. Hence option 2 will save the company money.

Testing to normal and abnormal conditions is outside the scope of this book but may be the subject of a future publication.

Compliance and conformity – manufacturing controls

Now our problems begin because we have only demonstrated total compliance on one or two systems: all subsequent manufactured units must meet these same essential requirements.

Let us assume that you have just received a test report confirming that your product meets the requirements of EN 60950 – what does that really mean to you, your company and to your customers? If the tests were performed on one unit, then it means that that one unit complied with the standard.

Your customer will expect every unit that they purchase to comply with the standard – but how do we do this? The answer is simple – we make them all the same. But things change. What will your company do when a component becomes unobtainable? Will you stop production? Will you resubmit the product for testing?

The solution is to define all safety critical features of the product and to place these under special control. That will give the manufacturing group the freedom to change other areas of the product: yet it will ensure safety critical components or features are not changed without your knowledge and approval.

It is up to the designer to define a 'Product Safety Description' so that the manufacturing, quality assurance and other groups ensure that there is conformity between the safety features of the products. Because of this conformity, they will all comply with the safety standard.

7 A simple guide to designs and reviews

Purpose

The purpose of this chapter is to provide designers with a simple methodology for designing compliant equipment and for conducting a product safety review.

We must note that there is a trade-off between design and compliance. We can make design decisions that commit moderate material costs and reduce compliance costs. Alternatively we will usually find that as we design to reduce manufacturing costs the costs demonstrating compliance will usually increase. However, it will enable us to produce designs whose compliance costs should be greatly reduced and, in most cases, it will help to reduce project time scales by many weeks.

Warning – by following the methods detailed in the following text a great many compliance issues will be removed; this, however, does not guarantee that all aspects of the harmonized standard that has been applied will have been met. A thorough review of the chosen harmonized standard *must always be made and documented*.

The simple design guide?

Which should we do first, design or review?

The answer to this is, if we are not the ones to ask and to question the fundamental basics of our everyday lives, who will question them and how will we learn? For me the important thing about questioning what many others consider to be 'facts' is that only when I start to question them – to ask about their detail – why we do things in certain ways – do I truly start to understand them myself.

Why is it important to question?

As practising engineers we are used to success. No engineer worth his salt would begin a new design unless they were certain that the product would be successful. In the product safety world we are not geared to willing products to be successful, we are looking for the (hopefully) very rare events that could cause a hazard. To be successful in product safety we must develop a new way of thinking and analysing: this means we must question everything about the product. Its design, its components, its manufacture, and its testing. To learn to question is good but learning how to question these things effectively will be one of the most difficult new skills to learn.

Why is understanding important?

Throughout our lives we are constantly moving from darkness into light – we are always striving towards improvement for ourselves, our families, our careers and the work that we do. We will be making decisions that directly or indirectly affect the futures of everyone about us, so we have a vested interest in getting it right. If we have a decision there are two basic methods:

1. We question and analyse and with our knowledge and experience we guess what will be the best solution.
2. We throw a dice and save all that effort.

Writing this book helped me to focus upon the 'right' mind-set to have when questioning what our peers may consider immovable facts. I hope that in reading it, it will help you to find the same understanding, allow you to reduce your chances of getting it wrong and improve your chances of getting it right. Please remember though – there are no panaceas, no magic wands and no promises.

So which should come first – the design review or the design – I believe that they go hand in hand. That way we minimize the cost of getting it wrong and improve our chances and our careers – if you cannot be sure then play the percentages. Never be afraid to ask. There is no such thing and a stupid question: only a dumb answer.

Please note that there are many standards. Some are *very* similar: some are quite different – therefore it is imperative to refer to the appropriate standard. There will be some standards that will place more onerous demands upon us and there will be others that allow relaxation. The following treatment is intended to provide a general middle-of-the-road approach to product safety design.

The basics

Hazards are not accidents without people

The operator

The standards allow us to assume differing levels of knowledge and competence for those handling our products. This allows us to design specific physical areas of a product for two types of people: the operator and the service engineer. The operator is assumed to be oblivious to electrical hazards, but does not act intentionally to create a hazard. Therefore they must be protected not only from hazards, but also from potential hazards. For example, mains cables have insulated wires covered by a jacket. This jacket provides a second layer of insulation in the event that the insulation of the wires may fail: *it is not acceptable to allow an operator to touch the brown or blue insulation of a mains.* Ensure that the word operator does not relate to any written (or other) instruction intended for service personnel.

Service personnel

It is assumed that service personnel will be reasonably careful in dealing with obvious hazards, but the design should protect against mishap by the use of warning labels, shields for hazardous voltage terminals, segregation of Safety Extra Low Voltage circuits from hazardous voltages etc.

The use of the words 'reasonably careful' and 'obvious' are important. All service engineers can expect the designer to take reasonable care in designing a product. The designer must be conscious of the fact that service engineers will not have the same product familiarity as they have, nor the same physical characteristics – they may have long hair and wear an employee pass on a long chain around their neck. For example, it is reasonable to warn the service engineer:

- to isolate power before removing covers; if live parts will be exposed, before the cover is removed
- which parts of the equipment contain hazardous voltages;
- which parts of the equipment are hot;
- what action to take to discharge large capacitors.

Similarly, it is reasonable to a service engineer to assume, for instance, that:

- unmarked metalwork and components are safe to touch;
- fans hidden within the equipment will be fitted with finger-guards;
- metal watch-straps, glasses and necklaces will not need to be removed before servicing the equipment;
- any lifting operation will require only one person;
- mechanical parts will not move and cause injury;
- laser radiation is not present.

Power connections to equipment

Connection to the supply

If we can characterize mains equipment by the way they are connected to the supply, there are three types:

1. Detachable power cord.
2. Fixed power cord.
3. Permanently connected.

Detachable power cord

The power cord can be unplugged from the supply and also from the equipment. This has several advantages. There is no need for strain relief: because the power cord is not permanently fixed if the lead is pulled it will simply come out of the socket. When supplying equipment to a variety of countries we need simply change the power lead: this can reduce the number of options and model numbers of the equipment.

Fixed power cord

Here the power lead is permanently attached to the equipment, strain relief is necessary and there should be instructions for replacing the power cord by a competent electrician or service engineer.

Permanently connected

In this equipment the power lead will be permanently attached to the supply and also to the equipment. Provisions for strain relief are obviously necessary, one item which is frequently overlooked by the equipment designer is to specify the supply protection and isolation necessary for servicing for maintaining the equipment. This must be provided by the installer or the customer and must be specified in the installation instructions.

Wire colour code

The wire between the building and the equipment must comply with local or national wiring regulations in terms of the cross-sectional area of the conductor etc. It must also conform to the correct colour code.

	Pluggable equipment	Permanently connected equipment		
Live	Brown	Brown and	OR	Red and
Neutral	Blue and	Blue and		Black
Safety Earth Ground	Green/Yellow	Green/Yellow		Green/Yellow

To provide any other colours for mains connections is a *criminal offence* in most countries in the EC. To understand why European requirements do not compromise on our colour codes we look to the following table.

Function	UK Building Wiring	US Wiring Code
Live	Red	Black
Neutral	Black	White
Safety Earth Ground	Yellow/Green	Green

Once we see that if a European electrician was attempting to connect equipment with US colours they would 'naturally' attempt to connect 'Green' to Safety Earth Ground, 'Black' to Neutral and 'White' to Live and we can see the potential danger!

Internal wiring colours

There is an important difference between the requirements for internal and external wiring in respect of colours. The reason for this is to avoid the hazard of making the wrong connections. There are (generally) no restrictions on the wiring colour for internal connections, this is good news for designers as it allows us to use colours to differentiate between individual wires. There is, however, one restriction: the colours of internal wires must not cause confusion. Hence the phase-neutral and ground terminals must be clear and unambiguous however equipment is designed to be mounted.

Type of plugs

There are two types of plug for us to consider:

1. Type A plug.
2. Type B plug.

Type A plug

Our domestic plugs are type A. The characteristics that make the type A plugs special are:

1. They can be assessed by the operator.
2. They are 'push to make' and 'pull to break (disconnect)'.
3. During connection and disconnection it is not possible to contact hazardous parts using an IEC test finger.
4. The time taken to disconnect this plug and to touch the pins is 1 second.

(This is important where energy is stored and accessible, e.g. via the power cord, motors, filters and circuits where the mains is rectified and is stored by a capacitor.)

The earth connection within the plug is made by a single fixing and therefore the loss of the Safety Earth Ground is 'only' one of the failures that we must consider. This is the reason that the maximum allowable earth leakage current through any single type A component is restricted to 3.5 mA – this is considered, by the standards, to be a 'safety' current for operators to contact under single fault conditions.

UK plug. Within the UK the type A plug is polarized and contains a fuse in the Live feed. Hence overload and overcurrent protection are achieved with the use of a single fuse. Operator protection from live parts (i.e. contact with the pins as the plug is inserted) is achieved by reinforced insulation on part of the Live and Neutral pins as shown.

The most common fuses used in this plug are 3 A or 13 A, ceramic high breaking capacity and the base of the plugs are marked with the fuse rating. Figure 7.1 shows these details. One common non-error is to supply a UK plug with a 13 A fuse rating but with a 3 A fuse fitted: if the fuse is replaced with the incorrectly indicated value there may be a fire. Where the rating of the equipment is less than 3 A cable of 0.5 mm² is permitted for cables of less than 2 metres long.

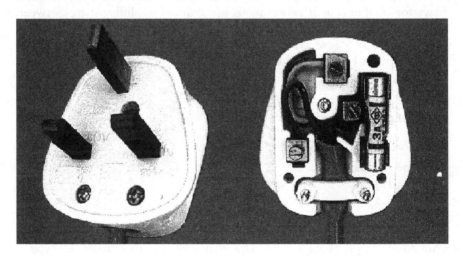

Figure 7.1   *13 amp mains plug*

Common European plugs. On mainland Europe the (DIN) Type A plug eliminates the possibility of finger contact, when the plug is partially inserted, by using a recessed section.

We need to be aware that there are three basic variations of these type A plugs (and sockets) for us to consider; the important thing to remember is that we can specify only plugs that will be suitable for all of three types of sockets, but if we specify the wrong type we can loose our Safety Earth Ground and create a hazard.

The DIN 49440 mains socket has two pins (at 3 o'clock and 9 o'clock) which provide Live and Neutral connections. The Safety Earth Ground connection is made by two wiping contacts at 6 o'clock and 12 o'clock – as shown in Figure 7.2. This type is not polarized: therefore we must carefully consider if we need to fit one or two fuses for short-circuit protection.

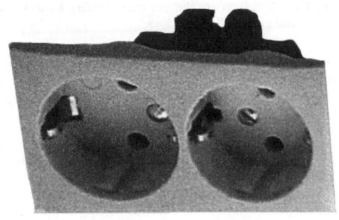

Figure 7.2   *DIN 49440 socket*

The second type is similar to, and physically interchangeable with, the earthed DIN 49440 but is designed for non-earthed, Class II, products only: the two wiping contacts are not fitted.

The third type of socket has a Safety Earth Ground pin that fits into the plug CEE7/V, (Figure 7.3). This gives the unwary design several opportunities for 'getting it wrong'.

Figure 7.3   *CEE7/V socket*

Several years ago I was reviewing a piece of Class I equipment. I found that it had been supplied with a power cord that was not compatible with the DIN 49440 socket. This meant that it would only provide a Safety Earth Ground to the equipment if connected to the CEE7/V style – the one with the earth pin in the socket. Unfortunately it was being supplied to a region of Europe where all sockets were DIN 49440. Hence 'as supplied' the Class II equipment did not have a Safety Earth Ground, was non-compliant and potentially hazardous.

US and other mains connectors in Europe. It may be useful to make a note about the use of the North American plug within Europe. Its use is illegal and expressly prohibited where the equipment is supplied to members of the public. This is because it is possible to touch the pins as they are inserted into the socket out to contact hazardous voltages (230 V). It may be acceptable to use the North American and similar styles of plug in a Service Access area – provided, of course, that they are rated for 250 V a.c. The designer would be well advised to fit a suitable warning in the area to warn of the 'unexpected' risk of electric shock.

Type B connectors

These are the industrial grade of connectors such as IEC 309. Many designers consider there use only in the industrial situation – this is not a true reflection of their usefulness to designers of low-power systems. They are not simple push-pull connectors but require some other operator to make or break them. This may be a mechanical interlock that must be lifted. Some type B connectors may require a twist-to-connect operation. The time taken to disconnect a type B connection and to touch its contacts is considered to be 5 seconds – this is useful if we have large capacitors that cannot be discharged within 1 second. The earth connection will have two independent fittings, hence a single failure will not disconnect the Safety Earth Ground.

The significance of this additional level of protection in the Safety Earth Ground circuit means that a single failure will not allow any Earth Leakage Current to pass through the operator (or service personnel). Therefore we are allowed to exceed the 3.5 mA limit (for Type A connectors), provided that our Safety Earth Ground wiring, joints and connections maintain the same reliability. Many of the harmonized standards permit Earth Leakage Currents of up to 5 per cent of the current taken by the equipment. (e.g. an Earth Leakage Current of 450 mA may be permitted for a 10 A load).

With the introduction of EMC regulations and the need to EMC filters the Earth Leakage Current of many products has increased significantly. (Try measuring the Earth Leakage Current taken by an EMC filter at a mains voltage of 255 Va.c. – this current will usually be much higher than the Earth Leakage Current of the equipment that it is filtering.)

To make matters worse these currents are cumulative, and so, when several items are connected through the same power lead the Earth Leakage Current can easily exceed the maximum limit for a type A plug of 3.5 mA. (The highest Earth Leakage Current that I have personally measured through a type A connector was 19.7 mA.)

Detachable power cord

The most common implementation of this is the IEC 320 connection used with computer and office equipment. This is a very attractive option for global export because it is possible for one design to satisfy any country requirements and, by simply supplying the appropriate power cord, the equipment can be supplied to other countries. Please note the previous comments about colour codes.

To modify a US power cable by wiring it to a European plug would be a criminal offence: the colours were not brown, blue and yellow/green or the socket fuse not rated for use at 230V. The use of <HAR> prevents the cable overcome with risks.

Do remember that the IEC 320 is a type A connection capable of a maximum of 3.5 mA earth leakage current. To provide a detachable power cord to equipment with an Earth Leakage Current of more than 3.5 mA requires a type B connection like the IEC 309 to be fitted to each end of the interconnecting cable. This section has covered many of the specific requirements of EN 60950 and other standards; however, the reader must confirm these requirements for themselves.

Fixed power cord

This method is commonly used for washing machines, irons, televisions, audio equipment and battery chargers.

The power cord is fixed to the equipment and emerges through a strain relief grommet and terminals in a type A or type B connection, in a 'harmonized' wire bearing the <HAR> mark. In many EU countries it is a legal requirement to fit a type A plug if equipment is intended for domestic use. Some countries allow power cords without plugs to be supplied provided they are for industrial use. If 'open-ended' power cords (without plugs) are to be supplied then the local regulations must be checked for each country of export.

Strain relief must prevent the wires from being pulled out, pushed in or twisted if a hazard can result. Some standards specify a minimum bend radius to prevent the power cord being damaged. Be aware of the mains plugs and sockets used in the country that you are supplying equipment to.

Permanently connected field wings terminal

These are achieved using a wire from a wall junction box or spur unit to the equipment. Common examples are cookers, hot water heaters, hand dryers, extractor fans, burglar alarm equipment and process equipment. As equipment providers we 'just' provide some wiring terminals for the customers' service personal to connect. However, in this case there are a number of issues, including use. Approved terminals or connection stripes will ensure compliance with other safety requirements.

We must clearly mark the required Safety Earth Ground, Neutral and Phase connections. There must be adequate space to make the terminations and to inspect the work afterwards. If the equipment can be mounted in various positions it may be necessary to fit two or more labels to identify the connections. We must provide or specify a disconnect device and overload protection.

'Overcurrent protection' may not present any difficulty but we must carefully consider short-circuit protection. For example, if we fit a 16 A circuit breaker with a 1500 A breaking capacity[1] this means that the circuit breaker will pass 16 A indefinitely. (Note for US and Canadian designers – IEC fuses and circuit breakers will pass the rated current, unlike UL and CSA rated overload devices which will break when their rated current is passed.) It will operate if the circuit is overloaded to, say, 20 A. However, if the customer installs the equipment so that short-circuit current could be in excess of 3000 A, due to being near the mains distribution panel etc., then we have a serious problem. Our (rated) 1500 A breaker may not break the full current and the wiring may overheat, catch fire etc.

We have a duty of care to warn installers of any special approaches needed to ensure a safe system of work for our equipment. Such warning should indicate:

- short-circuit protection required;
- high Earth Leakage Current;
- disconnection and 'lock-out' system maintenance;
- cooling requirements;
- special installation requirements;
- mechanical considerations – see page 165.

Equipment fusing

We now consider fuses, how many fuses, of what value and where to place them. There are few hard and fast rules other than we must protect against overcurrent and short-circuit faults.

If our internal mains wiring is covered by basic insulation then we must accept that a single fault can result in a short-circuit to Safety Earth Ground. If the equipment may be connected to a non-polarized mains supply (e.g. central European) then this fault condition can occur through either the internal Live or Neutral wiring: hence we must fit two fuses – one in the Live and the other in the Neutral. Fitting two fuses will require a warning label to be fitted, 'Double Pole/Neutral Fusing'. Many of these issues are overcome if we use one of the combined IEC 320 fused-switch-and-EMC filters.

Fuse rating is critical to correct operation and safety. Fuses should not blow when the equipment is energized – a useful test we can 'borrow' from the Canadian Standards Association (CSA) is to turn the equipment ON and OFF six times at intervals of 10 seconds. If the fuse blows then we can be sure that our customers will experience nuisance tripping. What message do we give if this happens: 'Fit a bigger fuse!!' Unless we want this to happen we must verify the correct fuse value. The next consideration is to ensure that the fuse is not too large to blow under likely fault conditions.

If we have an EMC filter rated at 6 A what value of fuse do we fit? Something up to 6 A but not greater.

What if we have a power supply with a 3 A internal fuse. If we fit 3 A fuses then we risk all of the 3 A fuses blowing together because there is no discrimination between them. Discrimination is the ability of the weakest fuse to blow without breaking (or weakening) other higher-rated fuses in the circuit. Determining the maximum value of the fuses is not easy. If we are lucky and have a single 'approved' power supply (PSU) then the documentation with the PSU will help. When we have multiple PSUs or a number of mains power items then choosing the correct fuse is not a simple matter and is well outside the scope of this introductory booklet.

Where fuses are operator accessible e.g. in an externally moulded mounted fuseholder or in one of the IEC type connectors, then we must specify clearly what type and value it should be replaced with. The information we are required to provide is voltage, current, and fusing characteristics.

Marking	*Meaning*
250 V 5A	250 volt, will pass 5 amp maximum steady state
250 V 5A T	250 volt, will pass 5 amp maximum steady state, time delay – slow blow
250 V 5A F	250 volt, will pass 5 amp maximum steady state, fast blow
250 V 5A F (H)	250 volt, will pass 5 amp maximum steady state, fast blow high breaking capacity

Rating label

As we are looking at the back of the equipment for labels now is the time to think about the rating label that we must fit. The minimum information that we must provide is:

Manufacturer's name	Enforcement offices will need this to progress any complaint
Address	
Voltage	This can be a range or a specific nominal voltage 115 V or 230 V are acceptable

Dual voltage or universal	115/230 V means it is selectable to operate on either voltage 115–230 V means that it will safely operate over the full voltage range
Frequency	50/60Hz means suitable for either frequency 50–60 Hz means it will operate over the full frequency range
Current	3A means that it will not take more than 3.3 amps
Duty cycle (if not stated the equipment is permanently rated)	10 s/5 min means a maximum 'on' time of 10 seconds with a repetition period of 5 minutes minimum

Product name or model number

Date of manufacture must be included – this may be month and year or serial number. This information is required so that defective patches can be identified. The CE Mark in letters not less than 5 mm high.

Choosing critical components

Please say to yourself 'I will always use "approved" critical components'. There is no option to this. If you chose a non-approved IEC connection then, in the words of many harmonized standards:

> 'Where safety is involved, components shall comply either with the requirements of this standard or with the safety aspects of the relevant IEC component standards – provided that both the component and its application are within the scope of that (relevant) IEC standard.'

If we follow this requirement and fit a component that does not have the safety mark of a European agency then we must conduct routine and regular type testing of the component and maintain manufacturing surveillance over the life of our product. Although there are many companies that choose not to specify 'approved components' they expose themselves to potential risk - because by fitting them, the manufacturer accepts full responsibility, for each component, in its application.

One can imagine the following scenario following a complaint:

Prosecution:	Do your safety components conform with IEC standards?
Manufacturer:	Of course they do.
Prosecution:	Do they have the mark of a European Safety Agency?
Manufacturer:	No.
Prosecution:	What testing do you conduct to verify that they comply with IEC standards?

Manufacturer:	Well we don't.		
Prosecution:	Then what actions have you taken in response to your obligation to take 'due diligence'?		
Manufacturer:	Errhmmm…		

Critical components will include any component connected to the primary supply or hazardous voltages and any component critical to safety. These components must be identified with an appropriate safety mark.

As with plastic parts, do not commit to any component until you physically hold a copy of the agency safety certificate. This will usually be a single sheet of paper and must be kept in the technical file for the product. This document will be vital when seeking a formal agency approval certificate for your product. A list of critical parts is given in the following table.

Critical items list

The following table identifies some typical Safety Critical Items. Note that where electrical safety is involved we select European component standards.

IEC or European component standard	UL94 flammability standard	Component	Possible operator-service warning
	94-V2	Air filter	
Y		Mains capacitors	Stored charge
Y		CRTs	Stored charge
Y		Circuit breakers	
Y		Conductive coatings	
Y		Connectors	
Y		Transformers and PSU	
Y	UL recognized	Fans above 30 V	
	UL recognized	Low power fans	
	94 VW1	Fibre optic cable	Eye damage
Y		Fuses and fuse holders	Replacement
Y		Safety switches	
Y		Line filters	
		Lithium batteries	Replacement – disposal instructions
Y		Mains connectors	
	UL94–various	All plastic parts	
Y		Power cords and mains cables	
Y		Mains voltage motors	
	UL94–V1	Printed circuit boards	
Y		Relays in safety applications or switching hazardous voltages	
Y		Products using primary power	

Y		Switches in safety applications or switching hazardous voltages	
Y		Transient voltage surge suppressors	
Y		Thermal controls	Min–Maximum
Y		External cables	
	UL94-VW1	Internal equipment wiring	

Earthing and grounding

Terminology

The first thing to avoid is confusion. By this I mean that when we talk about earthing we clearly distinguish between:

1. Signal ground.
2. EMC ground.
3. Safety Earth Ground.
4. Primary Earth Ground.
5. Secondary Earth Ground.

This discrimination is vital because Safety Earth Ground (SEG) connections are the only earth connections that can save lives!

Essential requirements of SEG

- All connections and joints are reliably made.
- SEG connections have no other function.
- A typical SEG connection will provide a minimum resistance of 0.1Ω at 25 A.
- Wire colours are (with exceptions for ribbon cable and braid) ~70%, Green ~30% Yellow stripe.

All connections and joints are reliably made

Compression joints are always made through metal: we *never* introduce plastic parts in the compression path. Figure 7.4 shows a non–compliant assembly. The plastic may age or soften and reduce the mechanical compression between the conductive parts and the SEG may fail.

Soldered joints *must* be made mechanically sound before soldering: a wire passed through a hole and soldered is not secure. Where soldered connection to an earth terminal must be made then it should be mechanically secured by fitting a suitable heatshrink sleeving to fix the wire to the terminal.

The wire must be crimped and a second crimp must be made to the insulation on the wire.

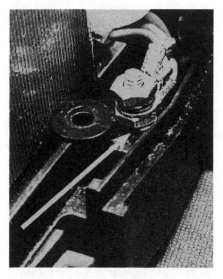

Figure 7.4 *In this example the Primary Earth Ground (arrowed) is compressed between a plastic surface – if the plastic is heated, or as it ages, the pressure between the grounding hardware may reduce and result in a high impedance under fault conditions*

Push-on connectors *must not* be the pull-off type. Those that have a small indentation to make disconnection difficult are not accepted by some agencies. My personal preference it to avoid this type of connector on earths at all costs. Figure 7.5 shows a reliable fixing.

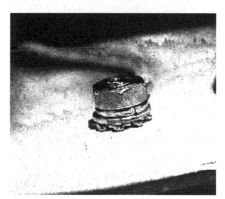

Figure 7.5 *A reliable Safety Earth Ground – note that a coarse 'Star' washer is used to cut through a paint finish. Constant pressure is applied onto the grounding hardware by a spring washer*

Safety Earth Ground (SEG) connections have no other function

If a conductive cover can come into contact with hazardous voltage covered only by basic insulation then its cover *must* be grounded (earthed) to SEG. That connection may be made through a flying lead to a terminal on the cover. Alternatively it may be made through a suitable (i.e. machine) screw into a tapped hole in the enclosure *provided* that this screw is not necessary to secure the cover.

We test this by removing the 'SEG screw' and carrying out all accessibility, mechanical and impact tests on the cover – if the cover remains fixed and prevents access then we have demonstrated that the 'SEG screw' only has one function. Then we remove *all* of the mechanical fixing and carry out the Earth Bond test – if the unit passes this test then we have demonstrated that the (single) 'SEG screw' provides an adequate SEG.

If we must joint two SEG wires then we must use a hook joint (Figure 7.6). This type of joint should also be used for all safety critical joints – including those in hazardous voltages.

Figure 7.6　*Hook joint*

Markings

Primary Earth Ground. There is only one Primary Earth Ground (PEG) in the equipment. This is the first point that the incoming SEG (from the supply) goes to. It is marked with the famous upside-down Christmas tree in a circle (Figure 7.7). There is an exception to this if we use a connector or terminal that is marked by the manufacturer with the Secondary Earth symbol.

Figure 7.7　*PEG marking*

Secondary Earth Ground. There may be any number of Secondary Earth Grounds. This will normally be taken from the Primary Earth Ground (PEG).

Electrochemical potential. If we take a non-rechargeable battery and allow it to discharge fully – and then leave it a long time – we notice that there will be a chemical discharge from it. These are the bi-products of the chemical reaction that produced the electrical energy. A similar thing will happen if we make electrical connections between parts of the wrong materials.

If we put zinc-plated steel into contact with nickel-plated steel we can measure a potential difference of about 0.9 V under the 'right' circumstances. After a few years of normal service (or after a few months in a warm humid atmosphere) the area between these materials will begin to look like a decomposing battery. The difference between them is, of course, the dead battery is unlikely to kill someone – whereas our (now) defective Safety Earth Ground might!

It is important to ensure that the materials we put in contact will generate only small electrochemical potentials. Less than 0.6 V is a typical maximum limit. These data are available in most standards – for example as Annex J of EN 60950 and the IEC 950 series of standards.

Earth Bond resistance

A typical SEG connection will provide a minimum resistance of $0.1\,\Omega$ at 25 A. This cannot be proven by calculation – we must test between appropriate points. This is usually done from the PEG (the IEC 320 inlet pin) to all operator accessible conductive parts that may contact basic insulation. (We do not mention operational insulation because we are forbidden to allow operational insulation to contact earthed parts.) This test should be repeated on 100 per cent of all production units for obvious reasons.

Wire colours

With the exception of specialist wire connection (such as 'ribbon' cable and braid) the required colours are ~70% Green, ~30% Yellow stripes. All wire of these colours will be considered as SEG – if any of the wiring or connections do not comply with the requirements for a reliable connection then the unit will probably fail to get an approval. Furthermore if the equipment is investigated by some third party it will be found non-compliant – this may be the more serious risk. The solution is to make all other earth wires some other colour – like green.

Switches, fuses, filters and internal wiring

These components should have an appropriate agency safety mark and approval, and be adequately rated. Remember that the fuse should have the lowest current rating of the three – this is because under overcurrent conditions it will protect the switch and other components. If the fuse does not provide this protection then there is a risk of fire.

Power supplies

This should also have an appropriate agency safety mark and approval. Ensure that there are no requirements for additional protection, and that this is correctly installed – in both Live and Neutral if necessary. Ensure that the precise operating conditions for the PSU are known and understood. Does it need a fire enclosure, does it need forced air cooling, does it need clearance for conventional cooling? Does it need to be mounted in a particular orientation?

The enclosure

General

First of all, forget about the enclosure as a small, neat compartment in which the electronic circuitry is housed – may have several of the following functions, and it is important to understand which *before* we go further:

1. To prevent the operator from reaching hazardous parts.
2. To prevent things falling in.
3. To prevent hazards from escaping.
4. To look pretty.

The enclosure we fit around our product may need to meet one or all of these requirements. Only one person will know this – the designer. It is therefore imperative that these requirements are understood and identified *before* any of the design is started. If certain components require a fire enclosure then we may be able to mount them together and reduce the cost of building the fire enclosure. Similarly other considerations can be taken into account that may eliminate the need for interlocks or unnecessary structures, earthing and insulation.

Operator access

The enclosure must prevent the operator gaining access to hazardous parts under normal circumstances. An enclosure with no holes will minimize considerations for operator access. It should ensure compliance with abnormal testing (ball impact and mechanical strength). It will comply with the flammability requirements for a fire enclosure, and meet the requirements of an electrical enclosure.

Enclosures that restrict operator accessibility and are mounted *outside* of the fire enclosure may have a flammability rating of UL94-HB or better.

Mechanical enclosure

The purpose of the mechanical enclosure is to protect both the equipment and the operator in the event of mechanical damage to the equipment. It protects the potentially hazardous elements within the equipment from mechanical impact and forces which may reasonably be applied during the life of the equipment. It prevents operator access to hazardous parts which may otherwise become exposed during mechanical impacts. There are tests such as the 'ball impact' which may be applied to the external services to verify that clearance distances are not compromised and the structure does not break and allow operator access.

Please note that sockets, switches and other similar components are exempt from these requirements provided that they have been verified to their relevant IEC component standard – this means that if they have an agency safety mark then we do not need to apply impacts to them. If the mechanical enclosure is mounted outside of the fire enclosure then it may have a flammability rating of UL94-HB or better.

Fire enclosure

If your enclosure is fixed (fastened to a wall or structure) or weighs more than 18 kg, then the fire enclosure shall have a flammability rating of 94 UL-5V. If you cut a hole to fit a plastic window in this structure, then the window that you fit must have a similar flammability rating of 94 UL-5V. If you cut holes for plastic fuses, switches etc., you need not consider their flammability, provided these components have an appropriate approval (from a body within the European Union). However, you must consider impact strength for these components as well as any plastic window within the enclosure.

If the weight of the unit is less than 18 kg, the required flammability rating for the fire enclosure is relaxed to UL94 V-1. All previous notes apply in terms of mechanical strength and flammability for plastic windows, etc. The simplest way to implement a fire enclosure is to identify the components which require one. These generally vary between standards but will include items and components such as transformers, solenoids, open contacts, any potential source of arcing, and all molten material. We now imagine a light source vertically over these components and the shadow that they would cast on the face and side of the enclosure.

We then imagine a 5° spread from these components and the area enclosed within this footprint will require a fire enclosure. This is the importance of grouping these components together – we avoid the need for producing a large fire enclosure. In its simplest form the fire enclosure is a sheet of metal without holes or gaps, it must be under the components and may form a barrier within the external enclosure. There are very special rules applying to many harmonized standards and these standards must always be consulted.

Decorative enclosure

This enclosure performs no safety function but 'just looks pretty!' We verify this by removing all parts of the decorative enclosure and verifying that none of the accessible parts presents a potential hazard. We will usually find that the electrical supply is provided by Safety Extra Low Voltage Energy Limited – and that the circuit has been analysed to identify any element that could produce an electrical, energy, fire or other hazard. If the decorative enclosure is mounted *outside* of the fire enclosure then it may have a flammability rating of UL94-HB or better.

Fastenings

Fastenings that can be removed to access hazardous parts must require the operator to use a tool, and if a tool is being supplied with the equipment, then these fastenings must require a different tool. If possible avoid using quarter turn or quick release fastenings. Should manufacturing tolerances, routine servicing, or ageing cause them to loosen during their expected life the retrofit costs can be substantial. There is now a wide variety of tamper-proof fastenings available – these are particularly useful where the product is going into the domestic market.

Flammability

This could be the subject of a book in its own right – hence the treatment here must be particularly superficial. All critical components in hazard voltages (switches, filters etc.) will have safety approval marks from independent agencies such as VDE, Nemko, Semko, Demko, TÜV, BSI etc. and these agencies will ensure that they have suitable voltage, current, mechanical and flammability ratings. Critical components in SELV (e.g. fans) may have a UL recognition mark (Figure 7.8) to show that they also have the correct flammability rating.

Parts going inside the fire enclosure must have a flammability of UL94-V2 or better - there will be exceptions to this in the standard that you choose to apply. Usually there is a requirement to material of UL94-V1 or better to be placed under parts that may emit arcs or molten material. If potential ignition sources are within 12 mm or so of plastic parts then those plastics must also, usually, have flammability ratings of UL94-V1 or better.

Figure 7.8

Labels

The designer *must* consider all single failures which could result in a hazard. The following quotation is from the General Safety Directive. I believe that there is a powerful relevance to parallel directives in the principles that they describe. Do remember that their purpose is to protect the consumer – there will, after all, be no prizes for adhering to the letter of safety directives while ignoring their spirit.

General Safety Directive Article 3-2: '...producers shall ... provide consumers with the relevant information to enable them to assess the risks inherent in a product throughout the normal or reasonably foreseeable period of its use, where such risks are not immediately obvious without adequate warnings, and to take precautions against those risks'.

From this we might think that we can protect operators from hazards with labels such as:

- Do not press this button.
- Do not open this door.
- Do not disconnect this plug.

This is not the case for a number of reasons.

General Safety Directive Article 3-2 continues: '...Provision of such warnings does not, however, exempt any person from compliance with the other requirements laid down in this directive ...'

As Article 2 puts it: '... the categories of consumers at serious risk when using the product, in particular children'.

So in response to the potential hazards above we must:

- Locate the button where the operator cannot reach it.
- Fit a lock to the door.
- Relocate the plug and socket, or incorporate an electrical safety interlock.

We can usually, however, warn service personnel of hazards and expect them to respond to our reasonable warning and advice. But what do you do when you see a warning that reads:

> Binnen dit gebied is UITSLUITEND de 'Safety Extra Low Voltage' bekabeling toegestaant.

It is not reasonable to expect all European service personnel to understand Dutch. We must translate the warning into the native language of the operator or user. Several of the harmonized standards identify English as a suitable language for service warnings. (It may be useful to note that English is acceptable worldwide for service warnings with one exception – Canada. In Canada service warnings *must be in English and French*.)

Finally the information of the warning, rating and other safety critical labels must be indelible – we usually check this by rubbing the surface with a petrol- and then a water-soaked cloth.

Examples

The following information may be useful in covering some of the more common labelling requirements for products.

Product label

Model number; type number; manufacture (and address); serial number (or date code).

Rating label – mains

Voltage; current; frequency.

Rating label – SELV

Voltage 'd.c.'; current; polarity.

Fuse – operator replaceable

Current; voltage; Fast, <blank for normal>, time delay; high breaking, <blank for normal glass 35A>

Fuse – service replaceable

F1, F2, F3, etc.

Include the details in the service documentation – this makes life very much easier.

Stability – service

Caution: High centre of gravity: This equipment must be supported until it is fixed into position.
    Caution: Unit may topple: To reduce risk of injury operate stabilizing mechanism before servicing.
    Do not leave unattended.

High Earth Leakage Current

Caution: High Earth Leakage Current: Earth connection essential before connecting supply.

Stored charge – service

Caution: Capacitor <add its circuit reference> stores a hazardous energy.

Disconnect device – service

Disconnect power <state method> before servicing.

Multiple power leads – service

Caution: This equipment is supplied by multiple power leads: Disconnect <state the number here> power leads before servicing.

Lamps

Caution: High pressure lamp may explode if handled incorrectly: Refer to service instruction.

Electrical hazard – service

IEC 3864 figure 5036 symbol (the lightning strike).

Electrical hazard – under tool removable panel

IEC 3864 figure 5036 symbol (the lightning strike).

Hot surface

IEC 417 figure 5041 symbol.

General – service

Refer to service instructions.

Lithium battery – service

Caution: Lithium battery: Danger of explosion if battery is incorrectly fitted: Replace only with same or equivalent type recommended by the manufacturer.

Battery – service disposal

Dispose of used batteries in accordance with manufacturer's instruction.

Batteries

Fuse batteries to comply with operator and service requirements. Typically if the output connections to a 10 V battery can deliver more than 100 A or 250 VA then it must be fused at 5 A. Check this carefully with the standard you are using.

Outlets and inputs and outputs

Usually we need to connect to the equipment. Power outlets will either be capable of providing the current for which they are rated (note that this also affects the equipment rating) or they will be labelled with the maximum current that can be drawn from them (this is usually the best idea). Beware of using UL style connectors in operator accessible areas because they will usually fail the accessibility tests as they are inserted into the socket.

Design all operator accessible inputs and outputs so that they can be short-circuited and so that they comply with the SELVEL requirements.

Fault conditions

Obviously if there is also a failure of basic insulation after an earth connection has failed then the enclosed and other operator accessible parts would be raised to 230 V. While this could, undeniably, be a potentially lethal situation it would require *two* independent failures – and the standards (generally) allow us to ignore this possibility.

It is important to understand that if one (or more) of the failures occurs as a consequence of the other then we consider that only a single failure has occurred. For example, a failure of basic insulation causes a high fault current that fuses a Safety Earth Ground connection without blowing the fuse – the enclosure becomes live. We consider this to be a single fault condition: where a single fault causes (several) consequential failures we consider the final potentially hazardous situation and any potentially hazardous situation that arises during the sequence of events.

To labour this point still further: if basic insulation failure causes a loss of Safety Earth Ground and then eventually a protection device isolates the supply we would consider the intermediary hazardous conditions as a non-compliance.

Internal cables and looms

Select 300 V UL recognized cable, this will meet the flammability requirements. Specify the appropriate temperature rating; 'T105' means rated for 105°C.

External cables and looms

See above. Check any sleeving material is UL recognized – this ensures that it has the correct flammability rating.

Mechanical consideration

Stability

There are detailed requirements and standards differ. Some of the requirements include 10° tilt, 250 N force, 800 N step. These are essential and no design should be started until the precise requirements are known – because they may affect the fundamental design of the product.

Strength

These also differ between standards. Some use a steel ball, others an impact hammer.

Accessibility

These also differ between standards. Some use the IEC finger (check which version – the standard has changed), force fingers, test pins, chains and other odd shapes. Check these before making *any* holes: it is most frustrating to find that they are ½ mm too big and must be changed!

Durability

Plastics creep, soften and crack. There are variety of tests called up in the standards. Mount push-fit connectors and switched into metal this can avoid some of the tests. Avoid mounting push-in, click-into-place parts into plastic mouldings.

Sharp edges

Deburr all metal edges – these can cut operators, service personnel and insulation.

Thermal

Ensure that the unit has adequate cooling, that critical components (and wiring) do not overheat, and that during normal and abnormal conditions the operator accessible surfaces do not become excessively hot. (You must refer to the relevant standard for this as there are significant differences between them.)

Fire

Choose materials with the highest flammability rating practicable. Often we can see similar parts (e.g. grommets) with flammability ratings from 'unknown' to UL94 5V (the highest) and with little difference in price. Wherever possible avoid putting plastic and flammable material near things that might (under normal and abnormal conditions) get hot. Avoid using any material inside the equipment whose flammability rating is less than UL94 V1. This will mean that you do not need to consider its proximity to hot parts or parts that may emit flaming particles, and will generally save a considerable amount of time.

For Class III (battery powered products) fit a small fuse or limit the energy (say to less than 15 W) – this can relax the need for a fire enclosure.

Chemical and ozone

This is a specialist area outside the scope of this book.

Radiation

This is a specialist area outside the scope of this book.

Note

1. A 1500 A breaking current means that the circuit breaker can interrupt a fault current of up to 1500 A. If the fault current is more than 1500 A the circuit breaker may not be capable of breaking it (usually because an arc is formed).

8 Simple guide to flammability requirements

When a fire enclosure is necessary

Does not require a fire enclosure	Requires a fire enclosure	EN 60950 ref	Notes
If the equipment can only be energized when an operator is present and a failure would be obvious to the operator.		4.4.5	If power cannot be applied if the *operator* is absent then none of the following apply and the equipment does not need a fire enclosure, but note 4.4.4 and Annex A.3. Ensure that this feature will not change, because the change could be very costly.
	All parts that could emit burning particles.	4.4.6 Note A	The purpose of *abnormal testing* is to demonstrate that components and assemblies do not emit burning particles. (Alternatively we may design a *fire enclosure* around the electrical parts and hence reduce the investigation and testing necessary – this can be cost effective for one-offs and small production runs.)
Impedance or thermally protected motors.		4.4.6	As with all Safety Critical Items: ensure that the agency test certificate is valid.
Motors that comply with EN 60950 Annex B.		4.4.5.2	Check if your supplier has suitable test evidence.
	All other motors.	4.4.5.1	Check that the agency test certificate is valid and covers the required voltage and insulation classification (Class I or II).
	Exposed (unenclosed) arcing parts.	4.4.5.1	Open-framed contactors or motors, relays with exposed contacts, uncovered fusible links – including pcb track designed to rupture.
	All parts within a limited power source (see 4.4.5.1 and 2.11).	4.4.5.1	By definition, the energy levels here are very high and components on the high energy side of the overcurrent protection receive special treatment.

Does not require a fire enclosure	Requires a fire enclosure	EN 60950 ref	Notes
Certain components supplied by a secondary circuit that also meets the requirements of a limited power circuit (see notes and 2.11).		4.4.5.2	The components must be mounted on material of flammability UL94V-1 minimum (e.g. a suitable pcb). The standard does not specify that the pcb must be below the components (and hence prevent burning material from falling). However, it will be prudent to check how your intended test house will interpret this requirement.
	Components having windings (e.g. transformers, relays, solenoids, inductors).	4.4.5.1	Some of these components may be contained and enclosed within their own fire enclosure. Check with your supplier and get a copy of the test certificate. Alternatively you may need to test the component 'in application'. Alternatively if abnormal testing demonstrates that there is not a hazard, there may be a relaxation depending upon the flammability rating of the materials involved.

Special requirement for any enclosure

Special requirement for any enclosure	Requires a fire enclosure	EN 60950 ref.	Notes
Exposed (unenclosed) arcing parts within 13 mm of the enclosure.	That local area of the enclosure must pass test in EN 60950 Annex A.3.	4.4.4	
Parts that could (under abnormal conditions) ignite the enclosure.	That local area of the enclosure must pass test in EN 60950 Annex A.4.	4.4.4	

Construction of the bottom of a fire enclosure

Feature	Condition	EN 60950 ref.	Notes
Openings of any size.	Outside the area swept by line C in EN 60950 Figure 11. May include components listed under 'Does not require a fire enclosure' column above.	4.4.6	Voids, non-flammable mechanical parts, and circuit elements contained in the 'Does not require a fire enclosure list' do not require a fire enclosure: however, flaming particles can be emitted within a 5°cone under all items that require the fire enclosure – unless proven otherwise by abnormal testing.
Openings of 40 mm² or less.	Not within the 5° cone extending from under any item that requires a fire enclosure. Allowable in areas under components of flammability UL94V-1, HF-1, or better.	4.4.6	Flaming particles are the greatest hazard category and take priority. See also Annexes A4 and A5 for arcing and high temperature parts.
Baffle construction.	See EN 60950 Figure 12. Acceptable within the 5° cone extending from under any item that requires a fire enclosure.	4.4.6	A single baffle is commonly used to provide access for power and signal cables (de-burr sharp edges!)
Metal mesh.	2 mm x 2 mm maximum pitch using wire of 0.45 mm diameter minimum. Acceptable within the 5° cone extending from under any item that requires a fire enclosure.	4.4.6	Using mesh is a very useful method of providing ventilation while minimizing the investigation required. NB. Any coating on the wire must be excluded from these dimensions. The dimensions given are absolute limits, they are not nominal values. Unless electrically bonded to the structure, this mesh must be investigated as an ungrounded conductive part.
Metal sheet with 'holes'.	Dimensions of the metal plate and holes must conform to the limits in Table 15. Acceptable within the 5° cone extending from under any item that requires a fire enclosure.	4.4.6	The dimensions given in Table 15 are absolute limits, they are not nominal values.
No openings.	Satisfies fire enclosure compliance requirement.	4.4.6	This is the easiest option in terms of compliance testing.

Feature	Condition	EN 60950 ref.	Notes
Mass of equipment less than 18 kg.	Fire enclosure must be of flammability rating UL94 V-1 or better.	4.4.4	If the equipment may be permanently fixed to a structure then it must be UL94 5V. If the equipment is nearly 18 kg consider the effects of future changes, additional features, alternative components. Consider the advantages of designing for UL94 5V.
Fixed equipment, or equipment with a mass of 18 kg or more.	Fire enclosure must be of flammability rating UL94 5V, or better	4.4.4	
Components that fill an aperture and form part of the fire enclosure.	Components must comply with the relevant IEC component standard.	4.4.4	Select components with relevant safety approvals.
Material for fire enclosure.	Metal, ceramic and glass (that is heat resistant, tempered, wired or laminated) does not require flammability testing.	4.4.4	
Exposed (unenclosed) arcing parts within 13 mm of the enclosure.	The local area of the fire enclosure must pass test in EN 60950 Annex A.3.	4.4.4	
Parts that could (under abnormal conditions) ignite the enclosure.	The local area of the fire enclosure must pass test in EN 60950 Annex A.4.	4.4.4	

Construction for the side of a fire enclosure

Feature	Condition	EN 60950 ref	Notes
Components that fill an aperture and form part of the fire enclosure.	Components must comply with the relevant IEC component standard.	4.4.4	Select components with relevant safety approvals.
Inside the area swept out by line 'C' in EN 60950 Figure 11.	The requirements are more severe.	4.3.16	It must be assumed that flaming parts may be emitted onto the sides of the fire enclosure.
Where the side of the fire enclosure is inside the area swept out by line 'C' in EN 60950 Figure 11. Wire mesh.	Metal mesh 2 mm x 2 mm maximum pitch using wire of 0.45 mm dia minimum. Acceptable within the 5° cone extending from under any item that requires a fire enclosure.	4.4.6	Using mesh is a very useful method of providing ventilation while minimizing the investigation required. NB. Any coating on the wire must be excluded from these dimensions. The dimensions given are absolute limits, they are not nominal values. Unless electrically bonded to the structure this mesh must be investigated as an ungrounded conductive part.
Where the side of the fire enclosure is inside the area swept out by line 'C' in EN 60950 Figure 11. Metal sheet with 'holes'.	Dimensions of the metal plate and holes must conform to the limits in Table 15. Acceptable within the 5° cone extending from under any item that requires a fire enclosure.	4.4.6	The dimensions given in Table 15 are absolute limits, they are not nominal values.
Where the side of the fire enclosure is inside the area swept out by line 'C' in EN 60950 Figure 11. No openings.	Satisfies fire enclosure compliance requirement.	4.4.6	This is the easiest option in terms of compliance testing.
Material for fire enclosure.	Metal, ceramic and glass (that is heat resistant, tempered, wired or laminated) does not require flammability testing.	4.4.4	

Feature	Condition	EN 60950 ref	Notes
Exposed (unenclosed) arcing parts within 13 mm of the enclosure.	The local area of the fire enclosure must pass test in EN 60950 Annex A.3.	4.4.4	
Parts that could (under abnormal conditions) ignite the enclosure.	The local area of the fire enclosure must pass test in EN 60950 Annex A.4.	4.4.4	
Outside the area swept out by line 'C' in EN 60950 Figure 11.	The requirements revert to the less severe requirements described in EN 60950 for the 'openings in the sides of fire enclosures'.	4.3.16	It must be assumed that flaming parts may be emitted onto the sides of the fire enclosure.
Openings of any size.	Where any bare part at hazardous voltage/ energy is not within the profile formed by Figure 10 volume 'V'.	4.3.16	
Outside the area swept out by line 'C' in EN 60950 Figure 11, and where there are bare hazardous parts within the profile defined by Figure 10.	Openings must have louvres (see Figure 9). Aperture must not exceed 5 mm in any dimension, or must not exceed 1 mm in width regardless of length.	4.3.16	The louvres are to prevent objects falling into the equipment, onto bare hazardous parts (primary, or supplies that are not limited energy sources, or that are energy hazards) and causing a fire. Apertures can be: less than 5.0 mm diameter or less than 1.0 mm of any length.
Outside the area swept out by line 'C' in EN 60950 Figure 11, and where there are no bare hazardous parts within the profile defined by Figure 10.	Not exceeding 5 mm in any dimension. Not exceeding 1 mm in width regardless of length. Have louvres (Figure 9).	4.3.16	Apertures can be: less than 5.0 mm diameter. or less than 1.0 mm of any length.
Mass of equipment less than 18 kg.	Enclosure must be of flammability rating UL94 V-1 or better.	4.4.4	If the equipment is nearly 18 kg consider the effects of future changes, additional features, alternative components. Consider the advantages of designing for UL94 5V.

Feature	Condition	EN 60950 ref	Notes
Mass of equipment is 18 kg or more.	Enclosure must be of flammability rating UL94 5V.	4.4.4	
Components fill an aperture and form part of the fire enclosure.	Components must comply with the relevant IEC component standard.	4.4.4	Select components with relevant safety approvals.
Material for fire enclosure.	Metal, ceramic and glass (that is heat resistant, tempered, wired or laminated) does not require flammability testing.	4.4.4	
Air filters.	See 'Components'.	4.4.3.6	
Where exposed (unenclosed) arcing parts within 13 mm of the enclosure.	The local area of the enclosure must pass test in EN 60950 Annex A.3.	4.4.4	
Where parts that could (under abnormal conditions) ignite the enclosure.	That local area of the enclosure must pass test in EN 60950 Annex A.4.	4.4.4	
Where fire enclosure is not required.		4.4.4	See 'When a fire enclosure is necessary' (above).

Construction for the top of a fire enclosure

Feature	Condition	EN 60950 ref	Notes
Components that fill an aperture and form part of the fire enclosure.	Components must comply with the relevant IEC component standard (but see 'fans' – components).	4.4.4	Select components with relevant safety approvals.
Material for fire enclosure.	Metal, ceramic and glass (that is heat resistant, tempered, wired or laminated) does not require flammability testing.	4.4.4	

Feature	Condition	EN 60950 ref	Notes
Exposed (unenclosed) arcing parts within 13 mm of the enclosure	The local area of the fire enclosure must pass test in EN 60950 Annex A.3.	4.4.4	
Parts that could (under abnormal conditions) ignite the enclosure.	The local area of the fire enclosure must pass test in EN 60950 Annex A.4.	4.4.4	
Where there are bare hazardous parts vertically beneath or within the 5° profile (similar to that shown in Figure 11).	Openings must have louvres (see Figure 8). Aperture must not exceed 5 mm in any dimension, or must not exceed 1 mm in width regardless of length.	4.3.15	An object dropped through a top opening may fall at an angle of 5° from the vertical. The louvres (defined in Figure 8) are to prevent objects falling into the equipment, onto bare hazardous parts (primary, or supplies that are not limited energy sources, or that are energy hazards) and causing a fire. Apertures can be: less than 5.0 mm diameter or less than 1.0 mm of any length.
Where there are no bare hazardous parts vertically beneath or within the 5° profile (similar to that shown in Figure 11).	There are no restrictions on the dimensions and construction of apertures in fire enclosure.	4.3.15	A chimney would meet these requirements provided that there were no bare parts at hazardous voltage. NB. Operator access must also be considered.
Mass of movable equipment less than 18 kg.	Enclosure must be of flammability rating UL94V-1 or better.	4.4.4	If the equipment is nearly 18 kg consider the effects of future changes, additional features, alternative components. Consider the advantages of designing for UL94 5V.
Equipment is fixed or has a mass of 18 kg or more.	Enclosure must be of flammability rating UL94 5V.	4.4.4	
Components fill an aperture and form part of the fire enclosure.	Components must comply with the relevant IEC component standard.	4.4.4	Select components with relevant safety approvals.

Material for fire enclosure.	4.4.4	Metal, ceramic and glass (that is heat resistant, tempered, wired or laminated) does not require flammability testing.
Air filters.		See 'Components'
Where exposed (unenclosed) arcing parts within 13 mm of the enclosure.	4.4.4	The local area of the enclosure must pass test in EN 60950 Annex A.3.
Where parts that could (under abnormal conditions) ignite the enclosure.	4.4.4	That local area of the enclosure must pass test in EN 60950 Annex A.4.
Openings of any size.	4.3.16	Where any bare part at hazardous voltage/energy is not within the profile formed by Figure 10 volume 'V'.
Where fire enclosure is not required	4.4.4	See 'When a fire enclosure is necessary'

Critical items list

The following table identifies some typical Safety Critical Items. Note that where electrical safety is involved we select European component standards.

IEC or European component standard	UL94 flammability standard	Component	Possible operator service warning
Y	94-V2	Air filter	
Y		Mains capacitors	Stored charge
Y		CRTs	Stored charge
Y		Circuit breakers	
Y		Conductive coatings	
Y		Connectors	
Y		Transformers and PSUs	
	UL recognized	Fans above 30 V	
	UL recognized	Low power fans	
Y	94 VW1	Fibre optic cable	Eye damage
Y		Fuses and fuse holders	Replacement
Y		Safety switches	
		Line filters	
Y		Lithium batteries	Replacement – disposal instructions
		Mains connectors	
Y	UL94-various	All plastic parts	
Y		Power cords and mains cables	
	UL94-V1	Mains voltage motors	
Y		Printed circuit boards	
		Relays in safety applications or switching hazardous voltages	
Y		Products using primary power	
Y		Switches in safety applications or switching hazardous voltages	
		Transient voltage surge suppressors	
Y		Thermal controls	Min – Maximum
Y	UL94-VW1	External cables	
Y		Internal equipment wiring	

9 Creepage and clearance

EN 60950 standard reference	Synopsis	Guidance notes
2.9	Creepage distances, clearances and distances through insulation.	As we work through this section we shall find that creepage distances will usually be greater than the corresponding clearance distances. The layout of this section has been changed to follow my personal preference of calculating creepage distance first. First we shall calculate the creepage distance, then we shall try our best to exceed this dimension when we calculate the clearance distance. If we can make the clearance distance greater than the creepage distance then it is likely that we have made an indirect assumption, or omitted a factor from a calculation, since we know that the clearance distance will usually be smaller than the creepage distance.
2.9.1	General	In all product safety matters our objective *must always* be to try to find the worst case. This is because other people's lives depend on us. We are making these calculations for the benefit of the user, *not* to make life easy for the designer.
	Clearances shall be dimensioned in accordance with Sub-clause 2.9.2. Creepage distances shall be dimensioned in accordance with Sub-clause 2.9.3. Distances through insulation shall be dimensioned in accordance with Sub-clause 2.9.4.	
	Clearance and electric strength requirements are based on the expected overvoltage transients which may enter the equipment from the mains supply. According to IEC Publication 664, the magnitude of these transients is determined by the normal supply voltage and the supply arrangements. The latter are categorized into four groups as Installation Categories I to IV (also known as Overvoltage Categories I to IV). This standard assumes Installation Category II at the equipment supply terminals.	Transients may be caused by switching inductive loads or (the most severe form) by lightning. These transients are higher near their source and become attenuated by the inductive and capacitive components, within the power distribution system and building wiring, as they propagate to adjacent equipment. This Standard considers that pluggable type A, pluggable type B or permanently connected equipment should be designed to meet the conditions defined by IEC 664 Installation Category II. All of the data within this section of the Standard is based on that assumption. No allowance is made for the effects of highly capacitive loads which can modify the Installation Category: this may

seem unfair but consider a single component failure, e.g. capacitors becoming open-circuited. In this event the circuitry would again be subjected to Installation Category II and operator safety may be compromised.

Consider a spark plug. Each time an engine fires the air gap between the electrodes is broken down and a current passes between them. Each time this happens there is little change in the voltage at which breakdown occurs.

This is similar to the situation which may occur where electrical breakdown occurs across an air gap between components, or between adjacent tracks on a printed circuit board.

Now consider the insulation between windings on a transformer. If the insulation is overstressed and breakdown occurs then a pin-hole is usually formed: if the energy passing through this pin-hole is sufficiently high then local burning and charring will occur. Once this process has happened the insulation at this point has been permanently weakened. This situation may occur to the enamelled installation of windings on transformers, solenoids etc. It may also occur between solid insulation separating the primary (hazardous) and secondary (SELV) circuits – in this situation it is obvious that this type of failure must be avoided at all costs.
In the case of reinforced insulation (where there is no Safety Earth Ground) then it goes double!

The design of solid insulation and clearances should be coordinated in such a way that if an incident overvoltage transient exceeds the limits of Installation Category II the solid insulation can withstand a higher voltage than the clearances.

The requirements given in sub-clause 2.9 are for insulation operating at frequencies up to 30 kHz. It is permitted to use the same requirements for insulation operating at frequencies over 30 kHz until additional data is available.

Interpolation is not permitted for creepage distances or clearances, except where explicitly stated.

For operational insulation, creepage distances and clearances smaller than those specified in Sub-clause 2.9 are permitted subject to the requirements of items (b) or (c) of Sub-clause 5.4.4.

All other things being equal it is better to provide adequate operational insulation in all conditions. However, the standard recognizes that this is not always practicable and allows the

EN 60950 standard reference	Synopsis	Guidance notes
		designer either to prove, by appropriate electric strength tests, that operational insulation is adequate for the purpose: or to demonstrate that there is no consequential flammability or electric shock hazard following a failure of operational insulation.
	If the creepage distance derived from Table 6 is less than the applicable clearance, then the dimension for clearance shall be used as the minimum creepage distance.	There may be occasions where the clearance distance derived from this standard will be greater than the corresponding creepage distance: these events are usually rare. However, the standard requires that the creepage distance should at all times be greater than, or at least equal to, the minimum clearance distances. *Before* taking this action check the assumptions and calculations – (based on personal experience) you are more likely to find an error. Hence creepage distance is *always* greater than or equal to clearance distance.
	The values for Pollution degree 1 are applicable to components and assemblies which are sealed so as to exclude dust and moisture (see Sub-clause 2.9.6).	When the Standard states 'sealed' it means sealed. It does not include enclosures that have close fitting sections through which air may flow freely. Sealed enclosures include EMC filters and some d.c. – d.c. converters.
	The values for Pollution degree 2 are generally applicable to equipment covered by the scope of this standard.	Before embarking on a design for Pollution degree 2 consider the following: Is the equipment likely to be used in a less clean environment? Is it likely to stand on the floor where dust may be drawn inside by cooling fans or by convection currents? If I were to design for Pollution degree 3 would the extra creepage and clearance distances have a significant impact upon my design? If they do not then it may be beneficial to design for Pollution degree 3 and avoid the possibility of having to redesign circuit elements if the housing or location changes.

The values for Pollution degree 3 are applicable where a local internal environment within the equipment is subject to conductive pollution or to dry non-conductive pollution which could become conductive due to expected condensation.

For all power systems, the mains supply voltage in Tables 3, 4 and 5 is the phase-to-neutral voltage.

The requirements given are for insulation operating at frequencies up to 30 kHz. They may be used for insulation operating at frequencies over 30 kHz until additional data is available.

The following conditions are applicable during the assessment for compliance in accordance with Sub-clauses 2.9.2 and 2.9.3.

Movable parts shall be placed in the most unfavourable position.

For equipment incorporating ordinary non-detachable power supply cords, creepage distance measurements shall be made with supply conductors of the largest cross-sectional area specified in Sub-clause 3.3.5, and also without conductors.

When measuring clearances and creepage distances from an enclosure of insulating material through a slot or operating in the enclosure, the accessible surface shall be conditioned to be conductive as if it were covered by metal foil wherever it can be touched by the test finger, Figure 19, page 119, applied without appreciable force. (See Figure F14 point B)

Pollution degree 3 includes such equipment as printers (paper dust), photocopy machines (paper dust and toner), and floor mounted equipment with cooling fans.
Remember that if air filters are fitted they must be cleaned, or replaced, and that abnormal testing should include blocking all fans, simulating blocked air filters.

If your experience with high frequency and high voltages suggest that clearance, creepage, clearance and distance through insulation should be greater than suggested by this standard then you should base your designs upon your practical experience. It is always safe to exceed the requirements of this standard: it is never a good idea to compromise these design limits.
Remember that if your decision is wrong then you may need to justify it before a court of law – if you are unable to convince them that your decision was correct you risk imprisonment!

Investigation must always include the largest conductor and positions of fixings (there may be occasions where clearance distance is compromised by rotating a nut or a build-up of tolerances).

This test attempts to simulate the effect of placing a hand over slots, there will be a tendency for the skin to penetrate the slot and reduce clearance (and possibly clearance creepage distance). Several years ago I investigated an incident where the operator received a static discharge when moving a VDU: in that instance the risk to the operator was dropping the monitor onto his foot, not electrical discharge.

EN 60950 standard reference	Synopsis	Guidance notes
2.9.3	Creepage distances.	As promised we are taking these out of sequence. You will also notice that creepage is 'easier' to calculate than clearance: this gives us a chance to look at the tables and to see how they work. This is particularly important if we do not use them often and are 'always on the learning curve'.
	Creepage distances shall be not less than the appropriate minimum values specified in Table 6 taking into account the relevant conditions specified under the table, subject to Sub-clause 2.91.	This is an obvious statement, and we should now look at our clearance distance calculations as though we are trying to exceed the (creepage) figure that we have just calculated. In all product safety matters our objective *must always* be to try to find the worst case. This is because people's lives depend on us and we are making these calculations for the benefit of the user, *not* to make life easy for the designer.
	Compliance is checked by measurement taking account of the figures in Appendix F and subject to the conditions in Sub-clause 2.9.1.	
	Clearances in primary circuits shall be dimensioned in accordance with Table 3 and Table 4. Clearances in secondary circuits shall be dimensioned in accordance with Table 5. The relevant conditions under the tables shall be taken into account.	There are two conditions and methods of calculating clearances. The first relates to primary and secondary circuits only – here we refer to Table 3 and Table 4. (REMEMBER THAT UNEARTHED CONDUCTIVE SURFACES ARE SECONDARY CIRCUITS.) The second condition relates to clearances in secondary circuits. These circuits will usually be in the secondary circuits of a transformer – the conditions are relaxed because the transformer filters voltage transients that occur and reduces spikes on the secondary circuits. Remember to use the correct method for primary circuits!
	NB. For clearances which are provided for compliance with 6.2.1.2, Table 5 applies. A transient rating of 1.5 kV peak should be assumed except where it is known that incoming transients will be suppressed, in which case the appropriate transient rating should be used.	Remember that suppression should be in series and fail-safe. Suppression devices placed across power rails may fail: if this would compromise the clearance distances to SELV then we should fail safe in our design and use the higher figure.

The values in the tables are the minimum values which shall be applied after taking account of manufacturing tolerances and deformation which can occur due to handling, shock and vibration likely to be encountered during manufacture, transport and normal use.

Please remember that manufacturing tolerances must be added to the distances calculated. When deciding upon tolerances always consider the financial cost of changing the design to accommodate relaxations, changes in process etc. Remember also that tolerances are a safety critical feature of the design and must be carefully reviewed.

The specified clearances are not applicable to the air gap between the contacts of thermostats, thermal cut-outs, overload protection devices, switches of micro-gap construction and similar components where the clearance varies with the contacts. For air gaps between the contacts of interlock switches the requirements of Sub-clause 2.8.6 are applicable.

This will generally be 3 mm (as specified is §2.6.2).

For primary circuits operating on nominal mains voltages up to 300 V, where the repetitive peak voltage in the circuit exceeds the peak value of the mains supply voltage, the minimum clearance is the sum of the following two values:
– the minimum clearance value from Table 3 for an insulation working voltage equal to the mains supply voltage;
– the appropriate additional clearance value from Table 4.

Table 3 plus Table 4.

The values in parentheses in Table 4 shall be used:
– when the values in parentheses in Table 3 are used in accordance with condition 3 of Table 3,

Note that the working voltages are *up to* the figures listed in Table 5.
The distances look small but they apply to Operational, Basic and Supplementary insulation.
To avoid misjudging the figure for reinforced insulation remember:
Basic + Supplementary = Reinforced

These reduced figures in parentheses may only be used if manufacturing is subject to a quality control programme (see Annex R of the Standard).
In practice this is acceptable for basic insulation and for products which consist wholly of Reinforced Insulation. It is, however, very difficult to implement this where there is a mixture of Basic and Reinforced insulation, because the dielectric strength tests for Reinforced insulation may destroy the Basic insulation within the product.

EN 60950 standard reference	Synopsis	Guidance notes
	– for operational insulation. The total clearances obtained by the use of Table 4 lie between the values required for homogeneous and inhomogeneous fields. As a result they may not assure conformance with the appropriate electric strength test in the case of fields which are substantially inhomogeneous.	A homogeneous field is said to exist between flat surfaces of different potential. If there are spikes, sharp edges or other features there will be a change to the uniformity of the field which may result in dielectric breakdown.
	Compliance is checked by measurement taking into account the figures in Appendix F, subject to conditions detailed in Sub-clause 2.9.1.	The examples in Annex F show the internationally recognized method of measuring creepage distances and clearance distances. They cannot be reproduced here in this document for copyright reasons. The clearance distance is the most direct route between two points. It may consist of a single straight line, or a series of straight lines around obstacles. The creepage distances can be visualized as the path left by a slug as it crawls over the surface, and will bypass holes, gaps or slots of up to: 0.25 mm for Pollution Degree 1 1.00 mm for Pollution Degree 2 1.50 mm for Pollution Degree 3
	If necessary, a force shall be applied to any point on internal parts and to the outside of conductive enclosures, in an endeavour to reduce the clearance while taking measurements. The force shall have a value of:	The clearance distances (even the 10 mm required for reinforced insulation – see condition 7) must not be compromised by this force. I personally feel that it is useful to see what form of dent can be created during the steel ball test and to use this data to allow for 'reasonably foreseeable abuse' when designing these clearances and airgaps.
	– 10 N for internal parts;	
	– 30 N for enclosures.	
	The force is applied to enclosures by means of a rigid test finger having outline dimensions as in Figure 19, page 239.	
	Circuits shall not be subject to internally generated transient overvoltages exceeding the appropriate value for the mains supply voltage and installation category.	Transient overvoltages which exceed the test voltages in Sub-clause 5.3.2 are outside the scope of this document. Check with IEC Publication 664 to determine the transient limits.

2.9.4	Distance through insulation NB. See also 3.1.5.	Please refer to these sections of the Standard and the guidance notes.
	Unless otherwise specified (see Sub-clauses 2.1.3, 2.9.5 and 3.1.5) distances through insulation shall be dimensioned according to working voltage and to application of the insulation (see Sub-clauses 2.2.7 and 2.2.6) as follows:	
	– for working voltages not exceeding 50 V (71 V peak or d.c.), there is no thickness requirement; – supplementary insulation shall have a minimum thickness of 0.4 mm;	But remember the flammability of plastic materials must be known and appropriate. Remember *minimum*. Although not stated explicitly this is commonly assumed to be for 230 V a.c. nominal.
	– reinforced insulation shall have a minimum thickness of 0.4 mm when not subject to any mechanical stress which, at nominal operating temperature, would be likely to lead to deformation or deterioration of the insulating material.	Ditto above.
	The above requirements are not applicable to insulation in thin sheet materials irrespective of its thickness provided that it is used within the equipment protective enclosure and is not subject to handling or abrasion during operator servicing and one of the following applies:	For thin sheet we carry out a basic insulation (or a reinforced) dielectric test on either a single sheet, or if three insulation systems are used, with *all* combinations of two insulation sheets.
	– supplementary insulation, at least two layers of material, each of which will pass the electric strength test for supplementary insulation; or	A failure of either sheet will not cause insulation to fail.
	– supplementary insulation comprises three layers of material for which all combinations of two layers together pass the electric strength test for supplementary insulation; or	A failure of any one sheet will not cause insulation to fail.
	– reinforced insulation comprises at least two layers of material, each of which will pass the electrical strength test for Reinforced insulation; or	A failure of either sheet will leave reinforced insulation in place.
	– reinforced insulation comprises three layers of insulation material for which all combinations of two layers together pass the electric strength test for reinforced insulation.	A failure of any one sheet will leave reinforced insulation in place.

EN 60950 standard reference	Synopsis	Guidance notes
	Enamel or other insulating coating on winding wire such as is normally used in transformer construction is not considered to be insulation in thin sheet material.	Enamelled wire and bare copper wire are treated the same with respect to safety.
2.9.5	For printed boards whose conductors are coated with a suitable coating material, the minimum separation distances of Table 7 are applicable to conductors before they are coated, subject to the following requirements.	*Minimum!*
	Either one or both conductive parts and at least 80 per cent of the distances over the surface between the conductive parts shall be coated. Between any two uncoated conductive parts and over the outside of the coating, the minimum distances in Table 3, 4 or 5 apply.	If not coated with a 'suitable' material then the normal creepage and clearance distances apply. The distance through insulation then 0.4 mm insulation is needed. Alternatively we can use the multiple layers as described in 2.9.4.
	The values in Table 7 shall be used only if manufacturing is subject to a formal quality control programme, an example of which is given in Annex R. In particular double and reinforced insulation shall be subject to 100 per cent electric strength testing.	Remember that mixed basic and reinforced Insulation systems will cause significant product test problems.
	In default of the above conditions, the requirements of Sub-clauses 2.9.2 and 2.9.3 shall apply.	
	The coating process, the coating material and the base material shall be such that uniform quality is assured and the separation distances under consideration are effectively protected.	Internationally recognized product safety agencies, such as Underwriter's Laboratories, operate a variety of acceptable manufacturing schemes…
	The coating material shall also be tested to the requirements of IEC Publication 112 (1979) for material group IIIa or IIIb, as defined in Table V, condition 3 of this Standard.	…and test methods. Consider the 'extra' cost of buying from a UL listed manufacturer; it could be cheaper; and if you intend to supply the product to the US it will probably be necessary.
	Compliance is checked by measurement taking into account Figures F12 and F13 in Appendix F and by the following series of tests.	Please refer to details in the Standard for this information.

Table 1 Minimum clearances for insulation in primary circuits and between primary and secondary circuits (mm)

Circuits subject to Installation Category II

Insulation working voltage (see Sub-clause 2.2.7)	V r.m.s. (sinusoidal)	Nominal mains supply voltage <150 V (Transient rating 1500 V)						Nominal mains supply voltage >150 V <300 V (Transient rating 2500 V)						Nominal mains supply voltage >300 V <600 V (Transient rating 4000 V)		
		Pollution degrees 1 & 2			Pollution degree 3			Pollution degrees 1 & 2			Pollution degree 3			Pollution degrees 1, 2 & 3		
V peak or d.c.V		Op	B/S	R	Op	B/S	R	Op	B/S	R	Op	B/S	R	Op	B/S	R
71	50	0.4	1.0 (0.7)	2.0 (1.4)	1.0	1.3 (1.0)	2.6 (2.0)	1.0	2.0 (1.7)	4.0 (3.4)	1.3	2.0 (1.7)	4.0 (3.4)	2.0	3.2 (3.0)	6.4 (6.0)
210	150	0.7	1.0 (0.7)	2.0 (1.4)	1.0	1.3 (1.0)	2.6 (2.0)	1.4	2.0 (1.7)	4.0 (3.4)	1.7	2.0 (1.7)	4.0 (3.4)	2.0	3.2 (3.0)	6.4 (6.0)
420	300	Op 1.7 B/S 2.0(1.7) R4.0(3.4)												2.5	3.2 (3.0)	6.4 (6.0)
840	600							Op 3.0 B/S 3.2(3.0) R6.4(6.0)								
1 400	1 000							Op/B/S4.2 R6.4								
2 800	2 000							Op/B/S/R 8.4								
7 000	5 000							Op/B/S/R 17.5								
9 800	7 000							Op/B/S/R 25								
14 000	10 000							Op/B/S/R 37								
28 000	20 000							Op/B/S/R 80								
42 000	30 000							Op/B/S/R 130								

Conditions applicable to Table 1

1. This table is applicable to equipment that will not be subject to transients exceeding Installation Category II according to IEC Publication 664. The appropriate transient voltage ratings are given in parentheses at the top of each nominal mains supply voltage column. Where higher transients are possible, additional protection might be necessary in the mains supply to the equipment or to the installation.

2. The values in the table are applicable to operational (Op), basic (B), supplementary (S) and reinforced (R) insulation.

3. The values in parentheses are applicable to basic, supplementary or reinforced insulation only if manufacturing is subject to a formal quality control programme, an example of which is given in Annex R. In particular, double and reinforced insulation shall be subject to 100 per cent electric strength testing.

4. For basic, supplementary and reinforced insulation, all parts of the primary circuit are assumed to be at not less than the nominal supply voltage with respect to earth.

5. For working voltages between 2800 V and 42 000 V peak or d.c. interpolation is permitted between the nearest two points, the calculated spacing being rounded up to the next higher 0.1 mm of increment.

6. For an air gap serving as reinforced insulation between a part at a hazardous voltage and an accessible conductive part of the enclosure of floor standing equipment or of the non-vertical top surface of desktop equipment the clearance shall be not less than 10 mm.

The specified clearances are not applicable to the air gap between the contacts of thermostats, thermal cut-outs, overload protection devices, switches of micro-gap construction and similar components where the clearance varies with the contacts. For air gaps between the contacts of interlock switches the requirements of Sub-clause 2.8.6 are applicable.

Table 2   Additional clearances for insulation in primary circuits with repetitive peak voltage exceeding the peak value of the mains supply voltage

| Nominal mains supply voltage 150 V | | Nominal mains supply voltage >150 V ≤300 V | Additional clearance mm | |
Pollution degrees 1 & 2 Maximum repetitive peak voltage V	Pollution degree 3 Maximum repetitive peak voltage V	Pollution degrees 1, 2 & 3 Maximum repetitive peak voltage V	Operational, basic or supplementary insulation	Reinforced insulation
210 (210)	210 (210)	420 (420)	0	0
298 (290)	294 (300)	493 (497)	0.1	0.2
386 (370)	379 (390)	567 (574)	0.2	0.4
474 (450)	463 (480)	640 (651)	0.3	0.6
562 (530)	547 (570)	713 (728)	0.4	0.8
650 (610)	632 (660)	787 (805)	0.5	1.0
738 (690)	716 (750)	860 (881)	0.6	1.2
826 (770)	800 (840)	933 (958)	0.7	1.4
914 (850)	–	1 006 (1 035)	0.8	1.6
1 002 (930)	–	1 080 (1 112)	0.9	1.8
1 090 (1010)	–	1 153 (1 189)	1.0	2.0
–	–	1 226 (1 266)	1.1	2.2
–	–	1 300 (1 343)	1.2	2.4
–	–	– (1 420)	1.3	2.6

Table 3 Minimum clearances in secondary circuits (mm)

Circuits subject to Installation Category 1 (see condition 5)

V peak or d.c.V	V r.m.s. (sinusoidal) V	150V (Transient 800V) Poll 1&2 Op	B/S	R	Poll 3 Op	B/S	R	>150V 300V (Transient 1500V) Poll 1&2 Op	B/S	R	Poll 3 Op	B/S	R	>300V 600V (Transient 2500V) Poll 1,2&3 Op	B/S	R	Not subjected to transient overvoltage (condition 4) Poll 1&2 only Op	B/S	R
71	50	0.4	0.7	1.4	1.0	1.3	2.6	0.7	1.0	2.0	1.0	1.3	2.6	1.7	2.0	4.0	0.4	0.4	0.8
140	100	0.6	(0.4)	(0.8)	1.0	(1.0)	(2.0)	0.7	(0.7)	(1.4)	1.0	(1.0)	(2.0)	1.7	(1.7)	(3.4)	0.6	0.7	1.4
210	150	0.6	0.7	1.4		1.3	2.6	0.7	1.0	2.0	1.0	1.3	2.6	1.7	2.0	4.0		(0.6)	(1.2)
280	200							Op1.1	B/S1.4(1.1)	R2.8(2.2)				1.7	2.0	4.0	1.1	1.1	2.2
420	300							Op1.6	B/S1.9(1.6)	R3.8(3.2)				1.7	2.0 (1.7)	4.0 (3.4)	1.4	1.4	2.8
700	500													Op/B/S2.5		R5.0			
840	600													Op/B/S3.5		R5.0			
1 400	1 000													Op/B/S4.5		R5.0			
2 800	2 000													Op/B/S/R 8.4					
7 000	5 000													Op/B/S/R 17.5					
9 800	7 000													Op/B/S/R 25					
14 000	10 000													Op/B/S/R 37					
28 000	20 000													Op/B/S/R 80					
42 000	30 000													Op/B/S/R130					

Conditions applicable to Table 3

1. The values in the table are applicable to Operational (Op), Basic (B), Supplementary (S) and Reinforced (R) insulation.

2. The values in parentheses are applicable to basic, supplementary or reinforced insulation only if manufacturing is subject to a formal quality control programme, an example of which is given in Annex R. In particular, double and reinforced insulation shall be subject to 100 per cent electric strength testing.

3. For voltages between 2800 V peak or d.c. and 42 000 V peak or d.c. interpolation may be used between the nearest two points, the calculated spacing being rounded up to the next higher 0.1 mm increment.

4. The values are applicable to d.c. secondary circuits which are reliably connected to earth and have capacitive filtering which limits the peak-to-peak ripple to 10 per cent of the d.c. voltage.

5. Secondary circuits will normally be Installation Category I when the primary is Installation Category II. However, a floating secondary circuit shall be subject to the requirements for primary circuits in Table III unless separated from primary circuits by an earthed metal screen.

6. External signal cables should be prevented from introducing into secondary circuits that exceeds the applicable transient overvoltage limit, where they might result in a hazard.

7. For an air gap serving as reinforced insulation between a part at a hazardous voltage and an accessible conductive part of the enclosure of floor standing equipment or of the non-vertical top surface of desk top equipment the clearance shall be not less than 10 mm.

Table 4 Minimum creepage distances (mm)

Working voltage up to and including V r.m.s. or d.c.	Operational, basic and supplementary insulation Pollution degree 1 Material group I, II, IIIa and IIIb	Pollution degree 2 Material group I, II, IIIa and IIIb		Pollution degree 3 Material group I, II, IIIa and IIIb	
50	Use the appropriate clearance from Table 1 or Table 2	0.6 0.9	1.2	1.5 1.7	1.9
100		0.7 1.0	1.4	1.8 2.0	2.2
125		0.8 1.1	1.5	1.9 2.1	2.4
150		0.8 1.1	1.6	2.0 2.2	2.5
200		1.0 1.4	2.0	2.5 2.8	3.2
250		1.3 1.8	2.5	3.2 3.6	4.0
300		1.6 2.2	3.2	4.0 4.5	5.0
400		2.0 2.8	4.0	5.0 5.6	6.3
600		3.2 4.5	6.3	8.0 9.6	10.0
1 000		5.0 7.1	10.0	12.5 14.0	16.0

Conditions applicable to Table 4

1. For Reinforced insulation, the values for creepage distances are twice the values in the table for basic insulation.

Remember that the figures are for basic insulation only.

2. If the creepage distance derived from Table 6 is less than the applicable clearance from Table 3 and Table 4 or from Table 5 as appropriate, then the value for that clearance shall be applied as the value for the minimum creepage distance.

(Note repeated from above. In all product safety matters our objective must always be to try to find the worst case. This is because people's lives depend on us and we are making these calculations for the benefit of the user, not to make life easy for the designer.)

3. Material group I   600 CTI (Comparative Tracking Index)
   Material group II   400 CTI 600
   Material group IIIb   100 CTI 175

The CTI rating refers to the value obtained in accordance with Method A, IEC Publication 112; Method for Determining the Comparative and the Proof Tracking Indices of Solid Insulating Materials under Moist Conditions.

Only if we are short of the CTI may we use these figures: if we have any doubt whatever in the material being used, the process, or what may happen in the future it is safest to use the worst case figure for material group IIIb.

4. Where the material group is not known, material group IIIb shall be assumed.

Assume the worst case.

5. For working voltages of 127 V, 208 V and 415 V, creepage distances corresponding to 125 V, 200 V and 400 V may be used.

Unless there are extremely good reasons for using these concessions my personal advice is to ignore them. The reasons for this are that these concessions are not compatible with some other standards (e.g. UL 1950) and it is possible that this allowance may be removed in the future.

6. It is permitted to use minimum creepage distances equal to the applicable clearances for glass, mica, ceramic or similar materials.

7. Linear Interpolation is permitted between the nearest two points, the calculated spacing being rounded to the next higher 0.1 mm increment.

Remember TOLERANCES

Table 5　Minimum separation distances for coated printed boards (mm)

Maximum working voltage V r.m.s. or d.c.	Operational, basic or supplementary insulation	Reinforced insulation
63	0.1	0.2
125	0.2	0.4
160	0.3	0.6
200	0.4	0.8
250	0.6	1.2
320	0.8	1.6
400	1.0	2.0
500	1.3	2.6
630	1.8	3.6
800	2.4	3.8
1 000	2.8	4.0
1 250	3.4	4.2
1 600	4.1	4.6
2 000	5.0	5.0
2 500	6.3	6.3
3 200	8.2	8.2
4 000	10.0	10.0
5 000	13.0	13.0
6 300	16.0	16.0
8 000	20.0	20.0
10 000	26.0	26.0
12 500	33.0	33.0
16 000	43.0	43.0
20 000	55.0	55.0
25 000	70.0	70.0
30 000	86.0	86.0

For working voltages between 2000 and 30 000 V it is permitted to use linear interpolation between the nearest two points. the calculated spacing being rounded up to the next higher 0.1 mm increment.

Creepage and clearance EXAMPLE 1:

1. Re-read §1.4.11 and §2.2.7
2. We now turn to Table 6 (Minimum creepage distances) and look at the working voltage column.
3. In this example we will calculate the CREEPAGE DISTANCE for a nominal 230 V a.c. supply and an internal working SELV supply of 35 V (remember that we can consider SELV and ELV to be zero for calculating clearance distance cf. §2.2.7).
4. For creepage we use the nominal value (§2.2.7 last paragraph) 230 V a.c. The SELV secondary voltage is considered to be zero volts (§2.2.7) and for simplicity we shall consider the creepage working voltage to be 250 V.
5. We write down these assumptions and highlight the '250 V row'.
6. Our next consideration will be the pollution degree for the components and circuit elements. We are designing a floor standing equipment which may be used in a carpeted office: hence it is reasonable to assume that there will be a build-up of material across conductive surfaces – we will therefore begin our design for Pollution Degree 3 and will review creepage and clearance distances if we find that there are engineering problems in achieving them.
7. We write down assumptions and highlight the 'Pollution degree 3 – column'.
8. We now consider the Comparative Tracking Index (CTI) of the material: we will be using a fibreglass printed circuit material and the manufacturer's Certificate of Conformity states that it has a CTI of between 100 and 175. We can therefore use is from group IIIb.
9. We write down these assumptions and highlight the figures relating to material group IIIb.

The minimum creepage distance is 4.000 mm. NB Tolerances to allow for manufacturing processes and ageing are added to this 4 mm.

The corollary to this is that if ageing or manufacturing tolerance reduce the creepage distance to 3.999 mm then the whole product fails to comply with this standard.

REMEMBER:

The cost of having to change these dimensions can be prohibitive. It may involve hundreds of man-hours and £'000s worth of scrap: it could result in a death, expensive litigation and imprisonment.

It is not just important to get the right answer – we must be prepared for our assumptions to stand the test of time.

10 Summary

The flammability rating of materials (particularly plastics) is dependent on their thickness. If you hold a match to a tree stump the result will be burnt fingers. If the tree stump is shaved to a very fine point then you may succeed in lighting it at the thin section. If the taper is sufficiently gradual then the flame will burn to thicker sections until the whole trunk is on fire. Similarly, thin sections of plastic are more flammable, and will burn more readily, than thicker sections. If the design requires a plastic part of flammability rating V-1 then the thinnest section of that plastic part must be rated for V-1.

As with critical components, never commit a plastic part to a design until there is evidence that the part meets the required flammability rating. Adequate evidence would be a drawing or sketch detailing the smallest dimension of the plastic part, the manufacturer of the plastic, and a copy of the underwriter's laboratory certificate (or listing) for that plastic. Manufacturing controls should be put in place to require and maintain copies of certificates of conformity for plastic parts. All internal wiring, (such as flexible single core equipment wire) should be marked as UL recognized components, VW-1, or AWM. Three core mains cable within the equipment should be marked with the <HAR> symbol.

Connection to supply

Mains power cables terminated with a domestic type plug must use harmonized <HAR> cord. NEMA plugs are not permitted in an operator area: they may be used in a field service area only. Central European plugs are not polarized and it is possible to interchange live and neutral connections. Therefore any overload and isolation components in the equipment live wiring must also be inserted in the equipment neutral wiring.

An equipment rack should only have one power cord. In exceptional circumstances where more than one power cord is employed (such as for redundancy reasons) then all cords shall be connected to the same supply phase and it shall be made clear by marking, or otherwise, which equipment is powered from which cord (there should be labels at each end of the power lead.)

Isolation of supply

To service equipment safely, it must be possible to disconnect the supply mains. Isolation must be possible for the entire system and should be possible for each item (otherwise the whole equipment must be powered down). The means of isolation may be any one of the following:

- An on/off switch with a 3 mm contact gap in the equipment or rack, and with 0 and 1 to identify off and on.
- An appliance coupler (such as an IEC 320 inlet connector).
- The input mains plug.
- A distribution panel with an isolating switch.

In all cases it shall be obvious by proximity or marking which disconnection device isolates what equipment.

Mechanical hazards

These should be eliminated by good design. Where this is not possible and where the hazard is unlikely to result in an injury then the addition of warning and caution labels may be acceptable. Mechanical hazards include:

- Sharp edges or corners.
- Fan blades or other rotating objects.
- Drive shafts or pulleys.
- Pinch-points caused by moving objects.

Earthing

All metalwork which may become live in the event of a single fault must be reliably bonded to earth. This means:

- The metalwork must be bonded with green/yellow wire.
- Incidental bonding by means of hinges or mounting screws is not sufficient – a dedicated wire or screw must be used.
- Connections should be proof against the effects of vibration – star washers or the like should be used.
- The resistance of this path must not exceed $0.1\,\Omega$ when measured at a current of 25 A. (This does not include the resistance of the power cord.)

However, the resistance from the supply end of the power cable to any part of the structure must not exceed 0.2 Ω.

Types of insulation

Operational Insulation – provides sufficient isolation between circuit elements to allow the equipment to function. It may not be touched by the operator or by trained field service personnel.

Basic Insulation – provides basic protection from electric shock. Basic Insulation should not be considered as fail-safe. It cannot be touched or accessed by an operator. Basic Insulation may be touched by trained field service personnel.

Supplementary Insulation – provides an additional level of protection from electric shock. It is always used in conjunction with Basic Insulation. Supplementary Insulation is not considered fail-safe.

Double Insulation – is Basic plus Supplementary Insulation. Although Basic and Supplementary Insulation are considered as liable to fail individually, Double Insulation is considered fail-safe. It may be touched by operators and field service personnel.

Reinforced Insulation – is a single insulation system which provides the same isolation as Double Insulation.

Protection against electric shock

The basis behind safety standards is that there should be at least two levels of protection between casual users and electrical hazards. Generally (check with the relevant standard) the user should not be exposed to:

- Voltage levels in excess of 42.4 V peak 60 V d.c.
- Currents in excess of 8 A or
- Powers in excess of 15 VA.
- Energies greater than 20 J.

A power supply meeting these requirements is called SELVEL (Safety Extra Low Voltage Energy Limited). The two required levels of protection may be provided by:

- Two independent insulation systems (Double Insulation).
- One Insulation system equivalent to Double Insulation (Reinforced Insulation).
- One insulation system plus a safety ground. For example, Basic Insulation between primary and secondary, and the secondary connected to safety earth. Any failure of Basic Insulation will thus cause a high primary fault current and operate the mains fuse or current trip.

Resistance to fire

This is achieved by good electrical design coupled with the use of the correct materials for the application. Power cables are classified to meet the IEC requirements (and identified by the mark <HAR>). Internal wires are classified (UL, CSA and VDE are acceptable) and are to be tied away from possible sources of fire, such as transformers, fans and power semiconductors.

Racks should have a solid metal base under any of the following components:

- Arcing parts, such as switch or relay contacts, commutators.
- Windings, such as in a motor, transformer, solenoid, or relay.
- Wiring.
- Semiconductor devices such as rectifiers, thyristors, thermistors, transistors.
- Devices such as resistors, capacitors, inductors.

Fire hazards

Three elements are needed to support a fire: fuel, heat, and oxygen. But for the fire to become a hazard a fourth element is needed. The fire must propagate.

We have little control over the atmosphere in which our equipment operates so we concentrate our effort on fuel content, heat, and the enclosure construction. The standards set limits for the maximum of each element in equipment. This is important for the user because it ensures a balanced means of fire prevention for similar equipment of differing manufacture.

Limits on fuel

For movable equipment having a total mass exceeding 18 kg, fire enclosures are considered to comply without test, if, in the smallest thickness used, the material is of flammability Class 5V or better.

For movable equipment having a total mass not exceeding 18 kg, fire enclosures are considered to comply without test if, in the smallest thickness used, the material is of flammability rating UL94-V1 or better.

Components shall be mounted on material of flammability UL94-V1 or better, and shall be separated from less fire-resistant material by at least 13 mm of air.

Air filter assemblies shall be constructed of materials of flammability class UL94-V2 or better, or HF-2 or better.

Decorative parts, mechanical and electrical; enclosures and parts of enclosures, if located externally to fire enclosures, shall be of flammability class HB or better. Small external decorative parts that would contribute negligible fuel to a fire, such as nameplates, mounting feet, key caps, knobs and the like, shall be exempt from this requirement.

There are also detailed requirements for flammable liquids etc. and the construction of openings in fire enclosures.

Integrated circuit packages, transistor packages, opto-coupler packages, capacitors and other small parts are exempt from the flammability Class V-2 requirement if the parts are mounted on material of flammability Class V-1 or better.

Limits on heat

The most extreme form of heat hazard is fire, but the two lesser effects, personal injury and insulation damage, are sometimes overlooked.

Three distinct limits exist:

Fire risk – In normal use, equipment and its component parts shall not attain excessive temperatures. If necessary, the effects of direct and indirect heating must be investigated and tests including the hot wire ignition test and the high current arcing test must be carried out.

Personal injury – The temperature rises of handles, knobs, grips and the like shall be determined for all parts which are gripped in normal use and, if of insulating material, to parts in contact with hot metal.

Insulation damage

The following are examples of permanent damage to insulation:

- Severe or prolonged smoking or flaming.
- Electrical or mechanical breakdown of any associated component part.
- Flaking, embrittlement or charring of insulation.

The winding insulation must not be damaged and winding temperatures must not exceed the maximum temperatures stated with the standard.

List of common standards and reference documents

Please note that many of the documents in this list will have been superseded by the time this book goes into publication. The following list of standards should be used as a guideline to assist in finding the current standard and its issue.

Reference number	Reference document and title
EN 60335	Safety of household and similar electrical appliances
EN 60730	Automatic electrical controls for household and similar
EN 61010	Safety requirements for electrical equipment for measurement, control and laboratory use
EN 41003	Particular safety requirements for equipment to be connected to telecommunication networks (see also EN 60950)
EN 50060	Power sources for manual arc welding with limited duty
EN 50063	Safety requirements for the construction and the installation of equipment for resistance welding and allied processes
EN 50065	Signalling on low voltage electrical installations in the frequency range 3–148.5 kHz
EN 50078	Torches and guns for arc welding
EN 50083	Cabled distribution systems for television and sound signals
EN 50084	Safety of household and similar electrical appliances – requirements for the connection of washing machines, dishwashers and tumble-dryers to the water mains
EN 50087	Safety of household and similar electrical appliances – particular requirements for bulk-milk coolers
EN 50091	Uninterruptible power systems
EN 60034	Rotating electrical machines
EN 60051	Direct acting indicating analogue electrical measuring instruments and their accessories
EN 60061	Lamp caps and holders together
EN 60065	Safety requirements for mains operated electronic and related apparatus for household and similar general use
EN 60081	Tubular fluorescent lamps for general lighting service
EN 60204	Electrical equipment of industrial machines
EN 60215	Safety requirements for radio transmitting equipment
EN 60335	Safety of household and similar electrical appliances: particular requirements for cooking ranges, cooking tables, ovens and similar appliances for household use, kitchen machines, skin and hair care, heat pump and de-humidifiers
EN 60564	D.c. bridges for measuring resistance
EN 60651	Sound level meters
EN 60695	Fire hazard testing: test methods
EN60730	Automatic electrical controls for household and similar use: particular requirements for electrically operated door locks, energy regulators, timers and time switches, automatic electrical burner control systems, thermal protectors
EN 60804	Integrating-averaging sound level meters
EN 60825	Safety of laser products
EN 60945	Marine navigational equipment – general requirements – methods of testing and required test results
EN 61010	Safety requirements for electrical equipment for measurement, control and laboratory use Part 2-31: Particular requirements for hand-held probe assemblies for electrical measurement and test
EN 61131	Programmable controllers
EN 60950	of information technology equipment, including electrical business equipment

Plate 1 Problem: *a soldered joint that is not made fast, mechanically, before soldering; this joint also compromises clearance distances for the unprotected mains circuit (the fuse is after the inlet socket).* Solution: *provide better information, e.g. photographs showing safety critical details and details of critical components*

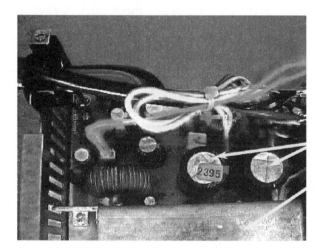

Plate 2 Problem: *the light-grey wire is rated 300 V and supplies the fan from SELV circuitry. This basic insulation does not provide adequate protection from the hazardous voltage on the bare metal of the mains socket; the capacitors are charged at up to 360 V peak and are storing 31 J; or the unearthed heat sink.* Solution: *fit supplementary insulation over the fan lead. Use wire with reinforced insulation (greater than 0.4 mm thick)*

Plate 3 Problem: *the connector 'A' fitted to the end of this flying lead is not physically retained with a clip. Once detached (a single fault) the flying lead can contact various hazardous parts 'B'.* Solution: *use a connector with an indent, or use cable tie or tie wrap to limit movement of flying lead, when it is unplugged*

Plate 4 *A European plug. The item has nine internationally recognized safety approvals. If the Safety Earth Ground corresponding socket has an earth 'PIN', then this plug is polarized. If the Safety Earth Ground connection is made via the arrowed contacts, then this plug is not-polarized. Therefore short-circuit protection between Neutral and Ground must be investigated and provided*

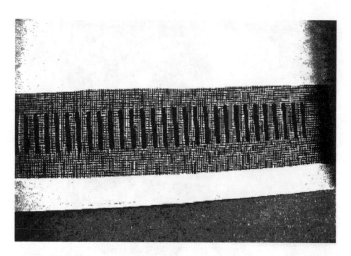

Plate 5 *Using mesh to provide a flame enclosure over large ventilation openings. Check: mesh is reliably fitted; adhesive is suitable for surface finish and highest temperature (test required); clearance distances provide reinforced insulation or clearance distances provide basic insulation and the mesh is reliably bonded to Safety Earth Ground*

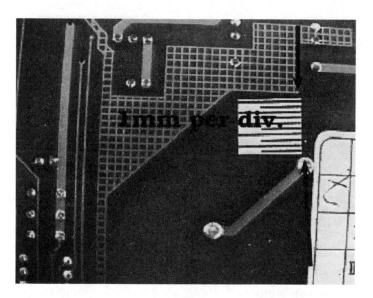

Plate 6 *Printed circuit board showing creepage distance. The creepage distance required is usually more than the corresponding clearance distance necessary to provide the same degree of insulation. If the creepage distance limits the circuit layout it is accepted practice to cut slots to eliminate the creepage path*

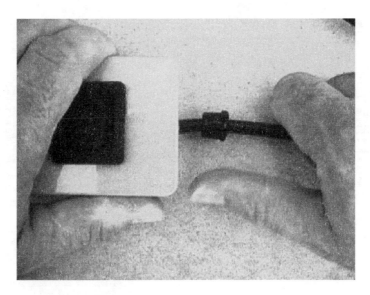

Plate 7 *An 'adapter box' failing the 'physical strength' test required in Section 3.2.5 of the standard*

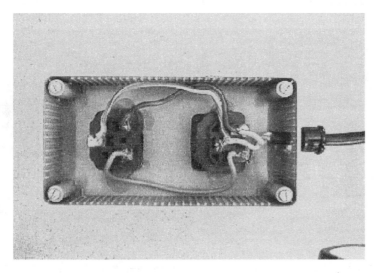

Plate 8 *This item was supplied as a sample. Non-compliances included: Safety Earth Ground connections are 'daisy-chained'; wires not bent and secured mechanically before soldering; operator may draw 20 A from these two outputs: the mains cable is rated for 6 A; the sockets were not reliably fixed; the enclosure failed the mechanical strength test and exposed hazardous parts; the enclosure also failed to comply with any of the UL94 flammability tests; there is no rating label or manufacturer's name*

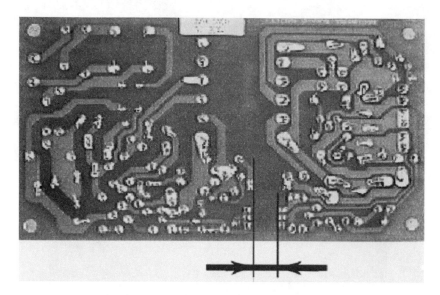

Plate 9 *'Approved' SELV power supply showing creepage and clearance distance, providing reinforced insulation, between hazardous and SELV components*

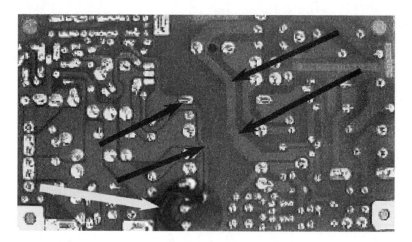

Plate 10 Problem: *'modified' SELV power supply. White arrow shows component added to 'SELV' circuitry. The possibility, and the effects, of this component becoming loose and contacting hazardous voltage would need to be explored. If changes are not carefully controlled then you may manufacture non-compliant products.* Solution: *define all safety critical aspects of the design with a Product Safety Descriptive Report – include photographs. (NB. See how the printed circuit tracks mirror one another to maintain the creepage distance)*

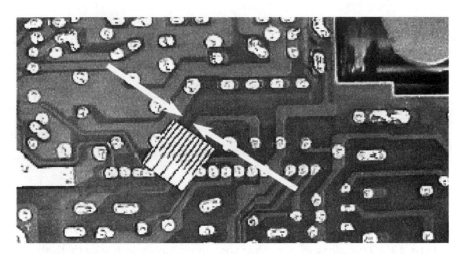

Plate 11 *Showing a power supply unit (PSU) that has basic insulation between hazardous and SELV circuits. This choice of PSU will, most likely, increase the amount of abnormal testing significantly*

Plate 12 *Attempting to measure the creepage and clearance distances can be difficult, as in the above case*

Plate 13 *The slots in the photograph do not comply with the requirements for a fire enclosure. Therefore the area encompassed by a line drawn 5° from the vertical from the top of the slots – this is the shaded area – must be a fire enclosure. If we mount the power supply on its side then the size of the fire enclosure will increase*

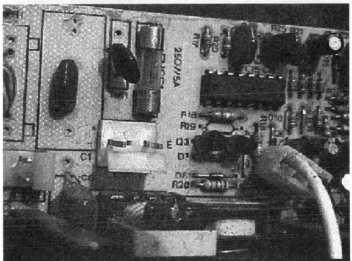

Plate 14   *The connector in these photographs is not positively located and can easily become unplugged allowing hazardous voltage to contact SELV (and therefore operator accessible) parts. The solution is to use a positively locking connector*

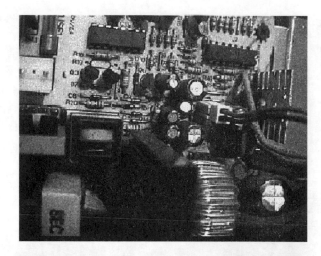

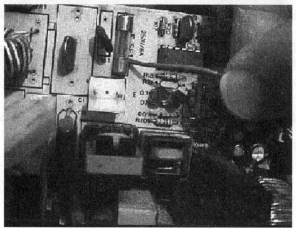

Plate 15   Problem: *in this example the wires of the connector in the top photograph are only connected by a single crimp. A single failure of either the crimp or the wire breaking will permit the (SELV) lead to contact hazardous voltage. Note that with only two crimps the wire will still result in the failure. Solutions: either use a crimp that crimps only the wire and has a second crimp that crimps onto the insulation or fit a tie wrap tightly around the wires to this connector so that if one of them breaks the other will hold it in position*

Plate 16 *These connections are 'daisy-chained' but because a single strand of wire is used the connections can be considered 'reliable' since if one solder joint fails the wire cannot move or break contact*

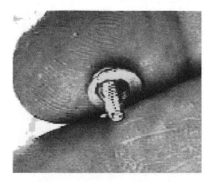

Plate 17 *This Safety Earth Ground fixing is a 'ring type' – which is good – but the fixing screw is much too small to provide a reliable connection*

Plate 18   *These two fuses illustrate the considerations and difficulty we can have in selecting components. The top fuse is a glass low breaking capacity fuse with a SEMKO approval. It is not rated for fault currents of more than 35 A – this current is equivalent to an earth fault of not less than 10 Ω – hence we must consider its use carefully in mains circuits. The lower fuse is a high breaking capacity fuse with no agency approvals. Technically this fuse is better in that it should break fault currents of over 1500 A – in other words it should be suitable for earth faults of not less than 0.24 Ω. Unfortunately without an IEC approval (e.g. an agency mark) we have no guarantee that this is so – and therefore to use this particular unapproved fuse may compromise the safety integrity of our product and our personal ability to demonstrate 'due diligence'*

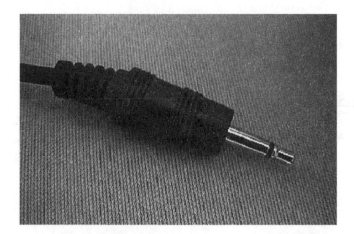

Plate 19   *This type of connector is often used to allow the 'operator' to connect d.c. power to equipment. Note that the electrical connectors are 'accessible' parts. This use is quite acceptable provided that the power is less than the maximum level defined by the Safety Extra Low Voltage Energy Limited requirement of the relevant standard – this is specified as 15 W maximum in several standards*

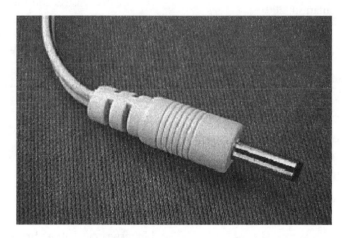

Plate 20   *In product safety terms this type of connector is identical to the illustration in Plate 19 as the electrical connections are 'operator' accessible. The power is less than the maximum level defined by the Safety Extra Low Voltage Energy Limited requirement of the relevant standard – this is specified as 15 W maximum in several standards*

Index